Problem-Solving Cases in Microsoft® Access and Excel, Second Annual Edition

Joseph A. Brady, Ph.D.

Ellen F. Monk

Australia • Canada • Mexico • Singapore • Spain • United Kingdom • United States

Problem-Solving Cases in Microsoft® Access and Excel,
Second Annual Edition

Executive Editor:
Mac Mendelsohn

Senior Product Manager:
Tricia Boyle

Development Editor:
DeVona Dors

Senior Marketing Manager:
Karen Seitz

Associate Product Manager:
Mirella Misiaszek

Editorial Assistant:
Amanda Piantedosi

Production Editor:
Kelly Robinson

Cover Designer:
Betsy Young

Cover Artist:
Rakefet Kenaan

Compositor:
GEX Publishing Services, Inc.

Manufacturing Coordinator:
Laura Burns

Printed in the United States of America

2 3 4 5 6 7 8 9 Glob 07 06 05 04

For more information, contact Course Technology, 25 Thomson Place, Boston, Massachusetts, 02210.

Or find us on the World Wide Web at: www.course.com

ISBN 0-619-21599-2

Dedications

Joe Brady: For Karen, again

Ellen Monk: For Peter, Caroline, and Catherine

Preface

For the past 14 years, we have taught MIS courses at the University of Delaware. From the start, we wanted to use good computer-based case studies for the database and decision support portions of our courses.

Prior to writing our first book of case studies, we could not find a textbook that met our needs. This surprised us because our requirements, we thought, were not unreasonable. First, we wanted cases that asked students to think about real-world business situations. Second, we wanted cases that provided students with hands-on experience, using the kind of software that they had learned to use in their computer literacy courses—and that they would later use in business. Third, we wanted cases that would strengthen students' ability to analyze a problem, examine alternative solutions, and implement a solution using software. Undeterred, we wrote our own cases.

After the success of our first book of cases (*Advanced Cases in MIS*, Course Technology, 2000), we wrote another one with new cases and improved tutorials (*Problem-Solving Cases in Microsoft Access and Excel, First Annual Edition*). That book was also a success, and now we have written the *Second Annual Edition*.

As with our prior casebooks, the tutorials prepare students for the cases, which are challenging but doable. The cases are organized in a way that helps the student think about the logic of the case and how to use the software to solve the business problem. The cases will fit well in an undergraduate MIS course, an MBA Information Systems course, or a Computer Science course devoted to business-oriented programming.

Book Organization

The book is organized into six parts:

1. Database Cases Using Access
2. Decision Support Cases Using Excel Scenario Manager
3. Decision Support Cases Using the Excel Solver
4. Decision Support Cases Using Basic Excel Functionality
5. Integration Case—Using Access and Excel
6. Presentation Skills

Part 1 begins with two tutorials that prepare students for the Access case studies. Parts 2 and 3 each begin with a tutorial that prepares students for the Excel case studies. All four tutorials provide students with hands-on practice using the software's more-advanced features—the kind of support that other books about Access and Excel do not contain. Part 4 asks students to use Excel's basic functionality for decision support. Part 5 challenges students to use both Access and Excel to find a solution to solve a business problem. Part 6 is a tutorial that hones students' skills in creating and delivering an oral presentation to business managers. The next section explores each of these parts in more depth.

Part 1: Database Cases

Using Access

This section begins with two tutorials and then presents five case studies.

Tutorial A: Database Design

This tutorial helps the student understand how to set up tables to create a database, without requiring the student to learn data normalization.

Tutorial B: Microsoft Access

The second tutorial teaches the student the more-advanced features of Access queries and reports—features that the student will need to know to complete the cases.

Cases 1-5

Five database cases follow Tutorials A and B. The student's job is to implement each case's database in Access so form, query, and report outputs can help management. The first case is a preliminary case, a "warm-up" case. The next four cases require a more demanding database design and implementation effort.

Part 2: Decision Support Cases

Using Excel *Scenario Manager*

This section has one tutorial and two-decision support cases requiring the use of Excel Scenario Manager.

Tutorial C: Building a Decision Support System in Excel

This section begins with a tutorial using Excel for decision support and spreadsheet design. Fundamental spreadsheet design concepts are taught. Instruction on the Scenario Manager, which can be used to organize the output of many "what-if" scenarios, is emphasized.

Cases 6-7

These two cases can be done with or without the Scenario Manager. In each case, the student must use Excel to model two or more solutions to a problem. The student then uses the outputs of the model to identify and document the preferred solution.

Part 3: Decision Support Cases

Using the Excel *Solver*

This section has one tutorial and two decision support cases requiring the use of the Excel Solver.

Tutorial D: Building a Decision Support System the Using Excel Solver

This section begins with a tutorial about using the Solver, which is a decision support tool for solving optimization problems.

Cases 8-9

Once again, in each case, the student uses Excel to analyze alternatives and identify the preferred solution—Using Basic Excel Functionality.

Part 4: Decision Support Cases

Using Basic Excel *Functionality*

Cases 10-11

The cases continue with two that use basic Excel functionality (i.e., the cases do not require the Scenario Manager or the Solver). Excel is used to test the student's analytical skills in "what if" analyses.

Part 5: Integration Case

Using Access and Excel

Case 12

The final case integrates Access and Excel. This case is included because there is a trend in industry to share data between multiple software packages to solve problems.

Part 6: Presentation Skills

Tutorial E: Giving an Oral Presentation

Because each case includes an assignment that gives students practice in making a presentation to management on the results of their analysis of the case, this section gives advice on how to create oral presentations. It also has technical information on charting and pivot tables, techniques that might be useful in case analyses or to support presentations. This tutorial will help students organize their recommendations, present their solutions in both words and graphics, and answer questions from the audience. For larger classes, instructors may wish to have students work in teams to create and deliver their presentations—which would model the "team" approach used by many corporations.

INDIVIDUAL CASE DESIGN

The format of the twelve cases follows this template.

- Each case begins with a *Preview* of what the case is about and an overview of the student's task.
- The next section, *Preparation*, tells students what they need to do or know to complete the case successfully. (Of course, our tutorials prepare students for the cases!)
- The third section, *Background*, provides the business context that frames the case. The background of each case models situations that require the kinds of thinking and analysis that students will need in the business world.
- This is followed by the *Assignment* sections, which are organized in a way that helps students to develop their analysis.
- The last section, *Deliverables*, lists what the student must hand in: printouts, a memorandum, a presentation, and files on disk. The list is similar to the kind of deliverables that a business manager might demand.

USING THE CASES

We have successfully used these cases, or cases very much like these, in our undergraduate MIS courses. We usually begin the semester with Access database instruction. We assign the Access database tutorials and then a case to each student. Then, for Excel DSS instruction, we do the same thing—assign a tutorial and then a case.

Technical Information

This textbook was quality-assurance tested using the Windows XP Professional operating system, and Microsoft Access 2003 or Microsoft Excel 2003 (depending on the case/tutorial).

Data Files and Solution Files

Case 12 requires a data file. This file can be found on the Course Technology Web site, where it is available to both students and instructors. Go to *www.course.com* and search for this textbook by title, author, or ISBN. You are granted a license to copy the data files to any computer or computer network used by individuals who have purchased this textbook.

Solutions to the material in the text are also available. These can also be found at *www.course.com*. Search for this textbook by title, author, or ISBN. The solutions are password protected and available only to instructors.

Instructor's Manual

An Instructor's Manual is available to accompany this text. The Instructor's Manual contains additional tools and information to help instructors successfully use this textbook. Items such as a Sample Syllabus, Teaching Tips, and Grading Guidelines are an example of the type of material that can be found in the Instructor's Manual. Instructors should go to *www.course.com* and search for this textbook by title, author, or ISBN. The Instructor's Manual is also password protected.

Acknowledgments

We would like to give many thanks to the team at Course Technology, including our Senior Product Manager, Tricia Boyle, and our Production Editor, Kelly Robinson. Special thanks goes to our Developmental Editor, DeVona Dors, whose attention to detail and valuable suggestions have helped to produce a polished and professional final product. As always, we acknowledge our students' diligent work.

Contents

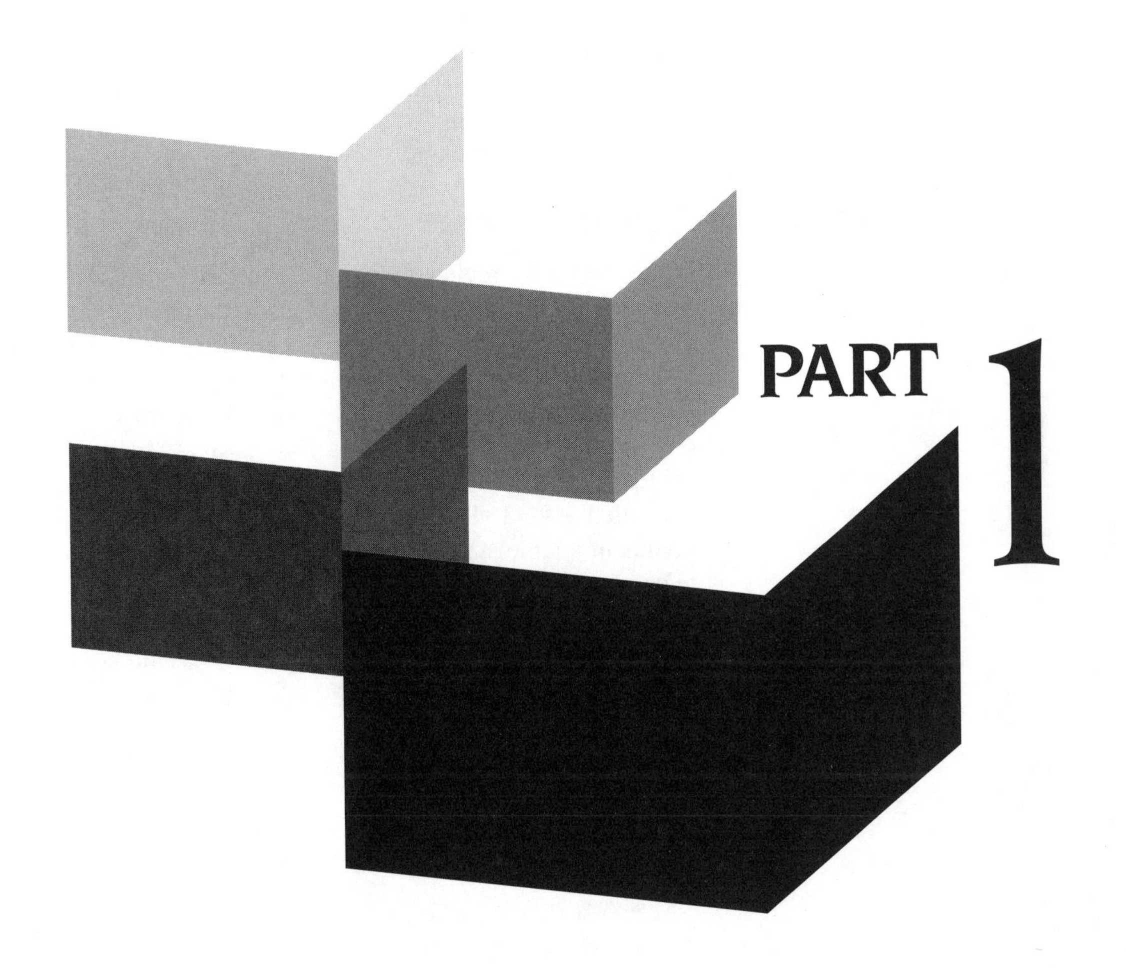

PART 1

Database Cases Using Access

Database Design

This tutorial has three sections. The first section briefly reviews basic database terminology. The second section teaches database design. The third section has a practice database design problem.

Review of Terminology

Let's begin by reviewing some basic terms that will be used throughout this textbook. In Access, a **database** is a group of related objects that are saved into one file. An Access **object** can be a table, a form, a query, or a report. You can identify an Access database **file** because it has the suffix **.mdb**.

A **table** consists of data that is arrayed in rows and columns. A **row** of data is called a **record**. A **column** of data is called a **field**. Thus, a record is a set of related fields. The fields in a table should be related to one another in some way. For example, a company might have employee data in a table called EMPLOYEE. That table would contain data fields about employees—their names, addresses, etc. It would not have data fields about the company's customers—that data would go into a CUSTOMER table.

A field's values have a **data type**. When a table is defined, the nature of each field's data is declared. Then, when data is entered, the database software knows how to interpret each entry. Data types in Access include the following:

- "Text" for words
- "Integer" for whole numbers
- "Double" for numbers that can have a decimal value
- "Currency" for numbers that should be treated as dollars and cents
- "Yes/No" for variables that can have only two values (1-0, on/off, yes/no)
- "Date/Time" for variables that are dates or times

Each database table should have a **primary key** field, a field in which each record has a *unique* value. For example, in an EMPLOYEE table, a field called SSN (for Social Security Number) could be a primary key, because each record's SSN value would be different from every other record's SSN value. Sometimes, a table does not have a single field whose values are all different. In that case, two or more fields are combined into a **compound primary key**. The combination of the fields' values is unique.

Database tables should be logically related to one another. For example, suppose that a company has an EMPLOYEE table with fields for SSN, Name, Address, and Telephone Number. For payroll purposes, the company would also have an HOURS WORKED table with a field that summarizes Labor Hours for individual employees. The relationship between the EMPLOYEE table and the HOURS WORKED table needs to be established in the database; otherwise, how could you tell which employees worked which hours? This is done by including the primary key field from the EMPLOYEE table (SSN) as a field in the HOURS WORKED table. In the HOURS WORKED table, the SSN field is then called a **foreign key**.

Data can be entered into a table directly or by entering the data into a **form**, which is based on the table. The form then inserts the data into the table.

A **query** is a question that is posed about data in a table (or tables). For example, a manager might want to know the names of employees who have worked for the company more than five years. A query could be designed to interrogate the EMPLOYEE table in that way. The query would be "run" and its output would answer the question.

A query may need to pull data from more than one table, so queries can be designed to interrogate more than one table at a time. In that case, the tables must first be connected by a **join** operation, which links tables on the values in a field that they have in common. The common field acts as a kind of "hinge" for the joined tables; the query generator treats the joined tables as one large table when running the query.

In Access, queries that answer a question are called **select** queries. Queries can be designed that will change data in records or delete entire records from a table. These are called **update** and **delete** queries, respectively.

Access has a **report** generator that can be used to format a table's data or a query's output.

DATABASE DESIGN

"Designing" a database refers to the process of determining which tables need to be in the database and the fields that need to be in each table. This section begins with a discussion of design concepts. The following key concepts are defined:

- Entities
- Relationships
- Attributes

This section then discusses database design rules, a series of steps we advise that you use to build a database.

Database Design Concepts

Computer scientists have formal ways of documenting a database's logic, but learning the notations and mechanics can be quite time-consuming and difficult. Doing this usually takes a good portion of a Systems Analysis and Design course. This tutorial will teach you database design by emphasizing practical business knowledge. This approach will let you design serviceable databases. Your instructor may add some more formal techniques.

A database models the logic of an organization's operation, so your first task is to understand that operation. You do that by talking to managers and workers, by observation, and/or by looking at business documents, such as sales records. Your goal is to identify the business' "entities" (sometimes called *objects*, in yet another use of this term). An **entity** is some thing or some event that the database will contain. Every entity has characteristics, called **attributes**, and **relationship**(s) to other entities. Let's take a closer look.

Entities

An entity is a tangible thing or event. The reason for identifying entities is that *an entity eventually becomes a table in the database*. Entities that are things are easy to identify. For example, consider a video store's database. The database would need to contain the names of videotapes and the names of customers who rent them, so you would have one entity named VIDEO and another named CUSTOMER.

By contrast, entities that are events can be more difficult to identify. This is probably because events cannot be seen, but they are no less real. In the video store example, one event would be the VIDEO RENTAL, and another would be HOURS WORKED by employees.

Your analysis is made easier by the knowledge that organizations usually have certain physical entities, such as:

- Employees
- Customers
- Inventory (Products)
- Suppliers

The database for most organizations would have a table for each of those entities. Your analysis is also made easier by the knowledge that organizations engage in transactions internally and with the outside world. These transactions are the subject of any accounting course, but most people can understand them from events in daily life. Consider the following examples:

- Organizations generate revenue from sales or interest earned. Revenue-generating transactions are event entities, called SALES, INTEREST, etc.
- Organizations incur expenses from paying hourly employees and purchasing materials from suppliers. HOURS WORKED and PURCHASES would be event entities in the databases of most organizations.

Thus, identifying entities is a matter of observing what happens in an organization. Your powers of observation are aided by knowing what entities exist in the database of most organizations.

Relationships

The analyst should consider the relationship of each entity to other entities. For each entity, the analyst should ask, "What is the relationship, if any, of this entity to every other entity identified?" Relationships can be expressed in English. For example, a college's database might have entities for STUDENT (containing data about each student), COURSE (containing data about each course), and SECTION (containing data about each section). A relationship between STUDENT and SECTION would be expressed as "Students enroll in Sections."

An analyst must also consider what is called the **cardinality** of any relationship. Cardinality can be one-to-one, one-to-many, or many-to-many. These are summarized as follows:

- In a one-to-one relationship, one instance of the first entity is related to just one instance of the second entity.
- In a one-to-many relationship, one instance of the first entity is related to many instances of the second entity, but only one instance of the second entity is related to an instance of the first.

- In a many-to-many relationship, one instance of the first entity is related to many instances of the second entity, and one instance of the second entity is related to many of the first.

To make this more concrete, again think about the college database having STUDENT, COURSE, and SECTION entities. A course, such as Accounting 101, can have more than one section: 01, 02, 03, 04, etc. Thus:

- The relationship between the entities COURSE and SECTION is one-to-many. Each course has many sections, but each section is for just one course.
- The relationship between STUDENT and SECTION is many-to-many. Each student can be in more than one section because each student can take more than one course. Each section has more than one student.

Worrying about relationships and their cardinalities may seem tedious to you now. However, you will see that this knowledge will help you determine the database tables needed (in the case of many-to-many relationships) and the fields that need to be shared between tables (in the case of one-to-many relationships).

Attributes

An attribute is a characteristic of an entity. You identify attributes of an entity because *attributes become a table's fields*. If an entity can be thought of as a noun, an attribute can be thought of as an adjective describing the noun. Continuing with the college database example, again think about the STUDENT entity. Students have names. Thus, Last Name would be an attribute, a field, of the STUDENT entity. First Name would be an attribute as well. The STUDENT entity would have an Address attribute, another field; and so on.

Sometimes, it is difficult to tell the difference between an attribute and an entity. One good way to differentiate them is to ask whether there can be more than one of the possible attribute for each entity. If more than one instance is possible, and you do not know in advance how many there will be, then it's an entity. For example, assume that a student could have two (but no more) Addresses—one for "home" and one for "on campus." You could specify attributes Address 1 and Address 2. On the other hand, what if the number of student addresses could not be stipulated in advance, but all addresses had to be recorded? You would not know how many fields to set aside in the STUDENT table for addresses. You would need a STUDENT ADDRESSES table, which could show any number of addresses for a student.

DATABASE DESIGN RULES

Your first task in database design is always to understand the logic of the business situation. You then build a database for the requirements of that situation. To create a context for learning about database design, let's first look at a hypothetical business operation and its database needs.

Example: The Talent Agency

Suppose that you have been asked to build a database for a talent agency. The agency books bands into nightclubs. The agent needs a database to keep track of the agency's transactions and to answer day-to-day questions. Many questions arise in running the business. For example, a club manager might want to know which bands are available on a certain date at a certain time or what fee the agent would charge for a certain band. Similarly, the agent might want to see a list of all band members and the instrument each plays, or a list of all the bands having three members.

Suppose that you have talked to the agent and have observed the agency's business operation. You conclude that your database would need to reflect the following facts:

1. A "booking" is an event in which a certain band plays in a particular club on a particular date, starting at a certain time, ending at a certain time, and for a specific fee. A band can play more than once a day. The Heartbreakers, for example, could play at the East End Cafe in the afternoon and then at the West End Cafe that night. For each booking, the club pays the talent agent, who keeps a five percent fee and then gives the rest to the band.
2. Each band has at least two members and an unlimited maximum number of members. The agent notes a telephone number of just one band member, which is used as the band's contact number. No two bands have the same name or telephone number.
3. No band members in any of the bands have the same name. For example, if there is a "Sally Smith" in one band, there is no Sally Smith in any other band.
4. The agent keeps track of just one instrument that each band member plays. "Vocals" is an instrument for this record-keeping purpose.
5. Each band has a desired fee. For example, the Lightmetal band might want $700 per booking and would expect the agent to try to get at least that amount for the band.
6. Each nightclub has a name, an address, and a contact person. That person has a telephone number that the agent uses to contact the club. No two clubs have the same name, contact person name, or telephone number. Each club has a target maximum fee. The contact person will try to get the agent to accept that amount for a band's appearance.
7. Some clubs will feed the band members for free, and others will not.

Before continuing, you might try to design the agency's database on your own. What are the entities? Recall that databases usually have CUSTOMER, EMPLOYEE, and INVENTORY entities and an entity for the revenue-generating transaction event. Each entity becomes a table in the database. What are the relationships between entities? For each entity, what are its attributes? These become the fields in each table. For each table, what is the primary key?

Six Database Design Rules

Assume that you have gathered information about the business situation in the talent agency example. Now you want to identify the tables for the database and then the fields in each table. To do that, observe the following six rules.

Rule 1: You do not need a table for the business itself. The database represents the entire business. Thus, in our example, Agent and Agency are not entities.

Rule 2: Identify the entities in the business description. Look for the things and events that the database must contain. These become tables in the database. Typically, certain entities are represented. In the talent agency example, you should be able to see these entities:

- *Things*: The product (inventory for sale) is Band. The customer is Club.
- *Events*: The revenue-generating transaction is Bookings.

You might ask yourself: Is there an EMPLOYEE entity? Also, isn't INSTRUMENT an entity? These issues will be discussed as the rules are explained.

Rule 3: Look for relationships between the entities. Look for one-to-many relationships between entities. The relationship between these entities must be established in tables, and this is done by using a foreign key. The mechanics of that is discussed in the next rule.

Look for many-to-many relationships between entities. In each of these relationships, there is the need for a third entity that associates the two entities in the relationship. Recall the STUDENT—Section many-to-many relationship example. A table is needed to show the ENROLLMENT of specific students in specific sections. (Enrollment can also be thought of as an event entity, and you might have already identified this entity. Forcing yourself to think about many-to-many relationships means that you will not miss it.)

Rule 4: Look for attributes of each entity, and designate a primary key. Think of entities as nouns. List the adjectives of the nouns. These are the attributes which, as was previously mentioned, become the table's fields. After you have identified fields for each table, designate one as the primary key field, if one field has unique values. Designate a compound primary key if no one field has unique values.

The attributes, or fields, of the BAND entity are Band Name, Band Phone Number, and Desired Fee. No two band names can be the same, it is assumed, so the primary key field in this case can be Band Name. Figure A-1 shows the BAND table and its fields: Band Name, Band Phone Number, and Desired Fee; the data type of each field is also shown.

Table Name: Band

Field	*Data Type*
Band Name (primary key)	Text
Band Phone Number	Text
Desired Fee	Currency

Figure A-1 The BAND table and its fields

Two BAND records are shown in Figure A-2.

Band Name (primary key)	*Band Phone Number*	*Desired Fee*
Heartbreakers	981 831 1765	$800
Lightmetal	981 831 2000	$700

Figure A-2 Two records in the BAND table

If there could be two bands called the Heartbreakers in the agency, then Band Name would not be a good primary key. Some other unique identifier would be needed. Such situations are common in business. Most businesses have many types of inventory, and duplicate names are possible. The typical solution is to assign a number to each product to be used as the primary key field. For example, a college could have more than one faculty member with the same name, so each faculty member would be assigned a Personal Identification Number (PIN). Similarly, banks assign a PIN for each depositor. Each automobile that a car manufacturer makes gets a unique Vehicle Identification Number (VIN). Most businesses assign a number to each sale, called an invoice number. (The next time you buy something at a grocery store, note the number on your receipt. It will be different from the number that the next person in line sees on their receipt).

At this point, you might ask why Band Member would not be an attribute of BAND. The answer is that you must record each band member, but you do not know in advance how many members will be in each band. Therefore, you do not know how many fields to allocate to the BAND table for members. Another way to think about Band Member is that they are, in effect, the agency's employees. Databases for organizations usually have an EMPLOYEE entity. Therefore, you should create a BAND MEMBER table with the attributes Member Name, Band Name, Instrument, and Phone. The BAND MEMBER table and its fields are shown in Figure A-3.

Table Name: Band Member

Field Name	*Data Type*
Member Name (primary key)	Text
Band Name (foreign key)	Text
Instrument	Text
Phone	Text

Figure A-3 The BAND MEMBER table and its fields

Five records in the BAND MEMBER table are shown in Figure A-4.

Member Name (primary key)	*Band Name*	*Instrument*	*Phone*
Pete Goff	Heartbreakers	Guitar	981 444 1111
Joe Goff	Heartbreakers	Vocals	981 444 1234
Sue Smith	Heartbreakers	Keyboard	981 555 1199
Joe Jackson	Lightmetal	Sax	981 888 1654
Sue Hoopes	Lightmetal	Piano	981 888 1765

Figure A-4 Records in the BAND MEMBER table

Instrument can be included as a field in the BAND MEMBER table, because the agent only records one for each band member. Instrument can thus be thought of as a way to describe a band member, much as the phone number is part of the description. Member Name can be the primary key because of the (somewhat arbitrary) assumption that no two members in any band have the same name. Alternately, Phone could be the primary key if it could be assumed that no two members share a telephone. Alternately, a band member ID number could be assigned to each person in each band, which would create a unique identifier for each band member handled by the agency.

You might ask why Band Name is included in the BAND MEMBER table. The common sense reason is that you did not include the Member Name in the BAND table. You must relate bands and members somewhere, and this is the place to do it.

Another way to think about this involves the cardinality of the relationship between BAND and BAND MEMBER. It is a one-to-many relationship: One band has many members, but each member is in just one band. You establish this kind of relationship in the database by using the primary key field of one table as a foreign key in the other. In BAND MEMBER, the foreign key Band Name is used to establish the relationship between the member and his or her band.

The attributes of the entity CLUB are Club Name, Address, Contact Name, Club Phone Number, Preferred Fee, and Feed Band? The table called CLUB can define the CLUB entity, as shown in Figure A-5.

Table Name: Club

Field Name	*Data Type*
Club Name (primary key)	Text
Address	Text
Contact Name	Text
Club Phone Number	Text
Preferred Fee	Currency
Feed Band?	Yes/No

Figure A-5 The CLUB table and its fields

Two records in the CLUB table are shown in Figure A-6.

Club Name (primary key)	*Address*	*Contact Name*	*Club Phone Number*	*Preferred Fee*	*Feed Band?*
East End	1 Duce St.	Al Pots	981 444 8877	$600	Yes
West End	99 Duce St.	Val Dots	981 555 0011	$650	No

Figure A-6 Club records

You might wonder why Bands Booked Into Club (or some such field name) is not an attribute of the CLUB table. There are two answers. First, you do not know in advance how many bookings a club will have, so the value cannot be an attribute. Furthermore, BOOKINGS is the agency's revenue-generating transaction, an event entity, and you need a table for that business transaction. Let us consider the booking transaction next.

You know that the talent agent books a certain band into a certain club on a certain date, for a certain fee, starting at a certain time, and ending at a certain time. From that information, you can see that the attributes of the BOOKINGS entity are Band Name, Club Name, Date, Start Time, End Time, and Fee. The BOOKINGS table and its fields are shown in Figure A-7.

Table Name: Bookings

Field Name	*Data Type*
Band Name	Text
Club Name	Text
Date	Date/Time
Start Time	Date/Time
End Time	Date/Time
Fee	Currency

Figure A-7 The BOOKINGS table and its fields—and no designation of primary key

Some records in the BOOKINGS table are shown in Figure A-8.

Band Name	*Club Name*	*Date*	*Start Time*	*End Time*	*Fee*
Heartbreakers	East End	11/21/05	19:00	23:30	$800
Heartbreakers	East End	11/22/05	19:00	23:30	$750
Heartbreakers	West End	11/28/05	13:00	18:00	$500
Lightmetal	East End	11/21/05	13:00	18:00	$700
Lightmetal	West End	11/22/05	13:00	18:00	$750

Figure A-8 Records in the BOOKINGS table

No single field is guaranteed to have unique values, because each band would be booked many times, and each club would be used many times. Further, each date and time could appear more than once. Thus, no one field can be the primary key.

If a table does not have a single primary key field, you can make a compound primary key whose field values together will be unique. Because one band can be in only one place at a time, one possible solution is to create a compound key consisting of the fields Band Name, Date, and Start Time. An alternative solution is to create a compound primary key consisting of the fields Club Name, Date, and Start Time.

A way to avoid having a compound key would be to create a field called Booking Number. Each booking would get its own unique number, similar to an invoice number.

Here is another way to think about this event entity: Over time, a band plays in many clubs, and each club hires many bands. The BAND-to-CLUB relationship is, thus, a many-to-many relationship. Such relationships signal the need for a table between the two entities in the relationship. Here, you need the BOOKINGS table between the BAND and CLUB tables.

Rule 5: Avoid data redundancy. You should not include extra (redundant) fields in a table. Doing this takes up extra disk space, and it leads to data entry errors, because the same value must be entered in multiple tables, and the chance of a keystroke error increases. In large databases, keeping track of multiple instances of the same data is nearly impossible, and contradictory data entries become a problem.

Consider this example: Why wouldn't Club Phone Number be in the BOOKINGS table as a field? After all, the agent might have to call about some last-minute change for a booking and could quickly look up the number in the BOOKINGS table. Assume that the BOOKINGS table had Booking Number as the primary key and Club Phone Number as a field. Figure A-9 shows the BOOKINGS table with the unnecessary field.

Table Name: Bookings

Field Name	*Data Type*
Booking Number (primary key)	Text
Band Name	Text
Club Name	Text
Club Phone Number	Text
Date	Date/Time
Start Time	Date/Time
End Time	Date/Time
Fee	Currency

Figure A-9 The BOOKINGS table with an unnecessary field—Club Phone Number

The fields Date, Start Time, End Time, and Fee logically depend on the Booking Number primary key—they help define the booking. Band Name and Club Name are foreign keys and are needed to establish the relationship between the tables BAND, CLUB, and BOOKINGS. But what about Club Phone Number? It is not defined by the Booking Number. It is defined by Club Name—*i.e., it's a function of the club, not of the booking.* Thus, the Club Phone Number field does not belong in the BOOKINGS table. It's already in the CLUB table, and if the agent needs it, he can look it up there.

Perhaps you can see the practical data entry problem with including Club Phone Number in BOOKINGS. Suppose that a club changed its contact phone number. The agent can easily change the number one time, in CLUB. But now the agent would need to remember the names of all the other tables that have that field as well, and change the values there too. Of course, with a small database, that might not be a difficult thing to recall. But in large databases having many redundant fields in many tables, this sort of maintenance becomes very difficult, which means that redundant data is often incorrect.

You might object, saying, "What about all those foreign keys? Aren't they redundant?" In a sense, they are. But they are needed to establish the relationship between one entity and another, as discussed previously.

Rule 6: Do not include a field if it can be calculated from other fields. A **calculated field** is made using the query generator. Thus, the agent's fee is not included in the BOOKINGS table because it can be calculated by query (here, five percent times the booking fee).

Practice Database Design Problem

Imagine this scenario. Your town has a library. The library wants to keep track of its business in a database, and you have been called in to build it. You talk to the town librarian, review the old paper-based records, and watch people use the library for a few days. You learn these things about the library:

1. Anyone who lives in the town can get a library card if they ask for one. The library considers each person who gets a card a "member" of the library.

2. The librarian wants to be able to contact members by telephone and by mail. She calls members if their books are overdue or when requested materials become available. She likes to mail a "thank you" note to each member on the yearly anniversary of their joining. Without a database, contacting members can be difficult to do efficiently; for example, there could be more than one member by the name of Sally Smith. Often, a parent and a child have the same first and last name, live at the same address, and share a phone.
3. The librarian tries to keep track of each member's reading "interests." When new books come in, the librarian alerts members whose interests match those books. For example, long-time member Sue Doaks is interested in Western novels, growing orchids, and baking bread. There must be some way to match such a reader's interests with available books. However, although the librarian wants to track all of a member's reading interests, she wants to classify each book as being in just one category of interest. For example, the classic gardening book *Orchids of France* would be classified as a book about orchids or a book about France, but not both.
4. The library stocks many books. Each book has a title and any number of authors. Conceivably, there could be more than one book in the library titled *History of the United States*. Similarly, there could be more than one author with the same name.
5. The library must be able to identify whether a book is checked out.
6. A member can check out any number of books in a visit. Conceivably, a member could visit the library more than once a day to check out books, and some members do just that.
7. All books that are checked out are due back in two weeks, no exceptions. The "late" fee is 50 cents per day late. The librarian would like to have an automated way of generating an overdue book list each day, so she could telephone the miscreants.
8. The library has a number of employees. Each employee has a job title. The librarian is paid a salary, but other employees are paid by the hour. Employees clock in and clock out each day. Assume that all employees work only one shift per day, and all are paid weekly. Pay is deposited directly into employees' checking accounts—no checks are hand-delivered. The database needs to include the librarian and all other employees.

Design the library's database, following the rules set forth in this tutorial. Your instructor will specify the format for your work.

Microsoft Access Tutorial

B TUTORIAL

Microsoft Access is a relational database package that runs on the Microsoft Windows operating system. This tutorial was prepared using Access 2003.

Before using this tutorial, you should know the fundamentals of Microsoft Access and know how to use Windows. This tutorial teaches you some advanced Access skills you'll need to do database case studies. This tutorial concludes with a discussion of common Access problems and how to solve them.

A preliminary caution: Always observe proper file-saving and closing procedures. Use these steps to exit from Access: (1) With your diskette in **drive A:**, use these commands: File—Close, then (2) File—Exit. This gets you back to Windows. Always end your work with these two steps. Never pull out your diskette and walk away with work remaining on the screen, or you will lose your work.

To begin this tutorial, you will create a new database called **Employee**.

At the Keyboard

Open a new database (in the Task Pane—New—Blank database). (According to Microsoft, the Task Pane is a universal remote control, which saves the user steps.) Call the database **Employee**. If you are saving to a floppy diskette, first select the drive (**A:**), and then enter the filename. **Employee.mdb** would be a good choice.

Your opening screen should resemble the screen shown in Figure B-1.

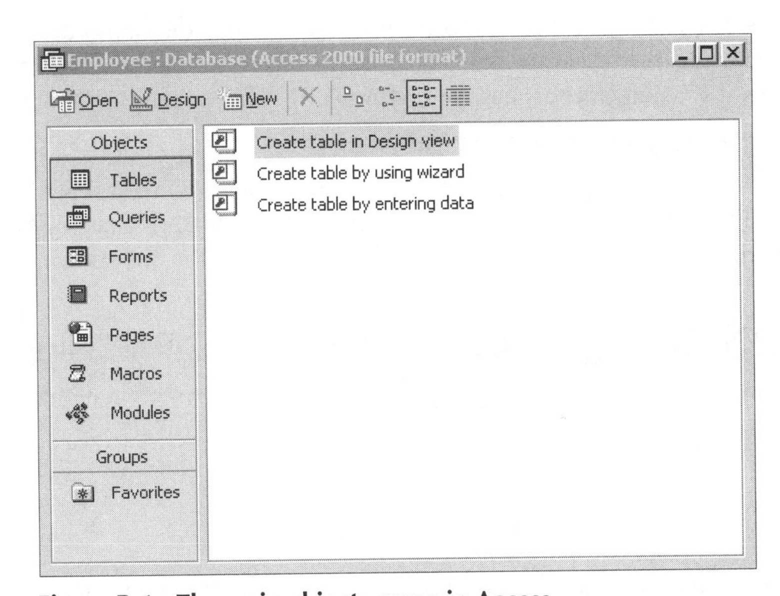

Figure B-1 The main objects menu in Access

In this tutorial, the screen shown in Figure B-1 is called the "main objects menu." From this screen, you can create or change objects.

CREATING TABLES

Your database will contain data about employees, their wage rates, and their hours worked.

Defining Tables

At the "main objects" menu, make three new tables, using the instructions that follow.

AT THE KEYBOARD

(1) Define a table called EMPLOYEE.

This table contains permanent data about employees. To create it, in the Table Objects screen, click New, then Design View, and then define the table EMPLOYEE. The table's fields are Last Name, First Name, SSN (Social Security Number), Street Address, City, State, Zip, Date Hired, and US Citizens. The field SSN is the primary key field. Change the length of text fields from the default 50 spaces to more appropriate lengths; for example, the field Last Name might be 30 spaces, and the Zip field might be 10 spaces. Your completed definition should resemble the one shown in Figure B-2.

Table1 : Table

	Field Name	Data Type	Description
	Last Name	Text	
	First Name	Text	
🔑	SSN	Text	
	Street Address	Text	
	City	Text	
	State	Text	
	Zip	Text	
	Date Hired	Date/Time	
	US Citizen	Yes/No	

Figure B-2 Fields in the EMPLOYEE table

When you're finished, choose File—Save. Enter the name desired for the table (here, EMPLOYEE). Make sure that you specify the name of the *table*, not the database itself. (Here, it is a coincidence that the EMPLOYEE table has the same name as its database file.)

(2) Define a table called WAGE DATA.

This table contains permanent data about employees and their wage rates. The table's fields are SSN, Wage Rate, and Salaried. The field SSN is the primary key field. Use the data types shown in Figure B-3. Your definition should resemble the one shown in Figure B-3.

Table1 : Table

	Field Name	Data Type	Description
🔑	SSN	Text	
	Wage Rate	Currency	
	Salaried?	Yes/No	

Figure B-3 Fields in the WAGE DATA table

Use File—Save to save the table definition. Name the table WAGE DATA.

(3) Define a table called HOURS WORKED.

The purpose of this table is to record the number of hours employees work each week in the year. The table's fields are SSN (text), Week # (number—long integer), and Hours (number—double). The SSN and Week# are the compound keys.

In the following example, the employee having SSN 089-65-9000 worked 40 hours in Week 1 of the year and 52 hours in Week 2.

SSN	Week #	Hours
089-65-9000	1	40
089-65-9000	2	52

Note that no single field can be the primary key field. Why? Notice that 089-65-9000 is an entry for each week. If the employee works each week of the year, at the end of the year, there will be 52 records with that value. Thus, SSN values will not distinguish records. However, no other single field can distinguish these records either, because other employees will have worked during the same week number, and some employees will have worked the same number of hours (40 would be common).

However, a table must have a primary key field. The solution? Use a compound primary key; that is, use values from more than one field. Here, the compound key to use consists of the field SSN plus the Week # field. Why? There is only *one* combination of SSN 089-65-9000 and Week# 1—those values *can occur in only one record*; therefore, the combination distinguishes that record from all others.

How do you set a compound key? The first step is to highlight the fields in the key. These must appear one after the other in the table definition screen. (Plan ahead for this format.) Alternately, you can highlight one field, hold down the Control key, and highlight the next.

AT THE KEYBOARD

For the HOURS WORKED table, click in the first field's left prefix area, hold the button down, then drag down to highlight names of all fields in the compound primary key. Your screen should resemble the one shown in Figure B-4.

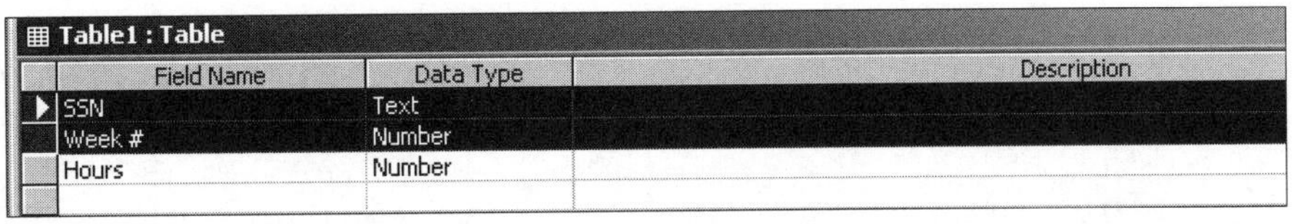

Table1 : Table

Field Name	Data Type	Description
SSN	Text	
Week #	Number	
Hours	Number	

Figure B-4 Selecting fields as the compound primary key for the HOURS WORKED table

Now, click the Key icon. Your screen should resemble the one shown in Figure B-5.

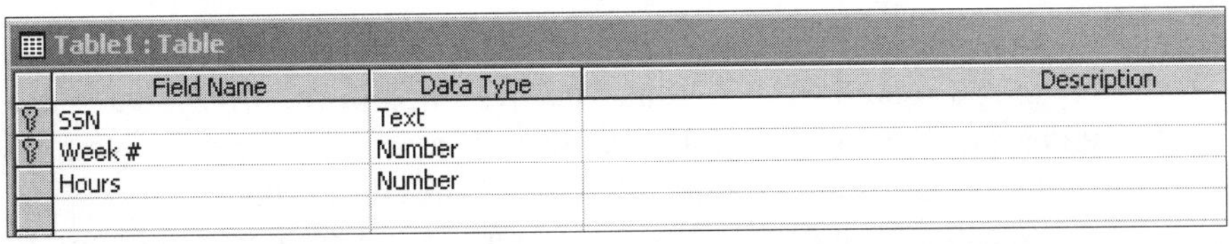

Table1 : Table

Field Name	Data Type	Description
SSN	Text	
Week #	Number	
Hours	Number	

Figure B-5 The compound primary key for the HOURS WORKED table

That completes the compound primary key and the table definition. Use File—Save to save the table as HOURS WORKED.

Adding Records to a Table

At this point, all you have done is to set up the skeletons of three tables. The tables have no data records yet. If you were to print out the tables, all you would see would be column headings (the field names). The most direct way to enter data into a table is to select the table, open it, and type the data directly into the cells.

At the Keyboard

At the main objects menu, select Tables, then EMPLOYEE. Then select Open. Your data entry screen should resemble the one shown in Figure B-6.

Employee : Table

Last Name	First Name	SSN	Street Address	City	State	Zip	Date Hired	US Citizen

Figure B-6 The data entry screen for the EMPLOYEE table

The table has many fields, and some of them may be off the screen, to the right. Scroll to see obscured fields. (Scrolling happens automatically as data is entered.) Figure B-6 has been adjusted to view all fields on one screen.

Type in your data, one field value at a time. Note that the first row is empty when you begin. Each time you finish a value, hit Enter, and the cursor will move to the next cell. After the last cell in a row, the cursor moves to the first cell of the next row, *and* Access automatically saves the record. (Thus, there is no File—Save step after entering data into a table.)

Dates (e.g., Date Hired) are entered as "6/15/04" (without the quotation marks). Access automatically expands the entry to the proper format in output.

Yes/No variables are clicked (checked) for Yes; otherwise (for No), the box is left blank. You can click the box from Yes to No, as if you were using a toggle switch.

If you make errors in data entry, click in the cell, backspace over the error, and type the correction.

Enter the data shown in Figure B-7 into the EMPLOYEE table.

Employee : Table

Last Name	First Name	SSN	Street Address	City	State	Zip	Date Hired	US Citizen
Smith	Albert	148-90-1234	44 Duce St	Odessa	DE	19722	7/15/1987	☑
Smith	John	123-45-6789	30 Elm St	Newark	DE	19711	6/1/1996	☑
Ruth	Billy	714-60-1927	1 Tater Dr	Baltimore	MD	20111	8/15/1999	☐
Jones	Sue	222-82-1122	18 Spruce St	Newark	DE	19716	7/15/2001	☐
Howard	Jane	114-11-2333	28 Sally Dr	Glasgow	DE	19702	8/1/2004	☑
Add	Your	Data	Here	Newark	MN	33776		☑

Figure B-7 Data for EMPLOYEE table

Note that the sixth record is *your* data record. The edit pencil in the left prefix area marks that record. Assume that you live in Newark, *Minnesota*, were hired on today's date (enter the date), and are a U.S. citizen. (Later in this tutorial, you will see that one entry is for the author's name and the SSN 099-11-3344 for this record.)

Open the WAGE DATA table and enter the data shown in Figure B-8 into the table.

Wage Data : Table

SSN	Wage Rate	Salaried
114-11-2333	$10.00	☐
123-45-6789	$0.00	☑
148-90-1234	$12.00	☐
222-82-1122	$0.00	☑
714-60-1927	$0.00	☑
Your SSN!	$8.00	☐

Figure B-8 Data for WAGE DATA table

Again, you must enter your SSN. Assume that you earn $8 an hour and are not salaried. (Note that Salaried = No implies someone is paid by the hour. Those who are salaried do not get paid by the hour, so their hourly rate is shown as 0.00.)

Open the HOURS WORKED table and enter the data shown in Figure B-9 into the table.

Hours Worked : Table

SSN	Week #	Hours
114-11-2333	1	40
114-11-2333	2	50
123-45-6789	1	40
123-45-6789	2	40
148-90-1234	1	38
148-90-1234	2	40
222-82-1122	1	40
222-82-1122	2	40
714-60-1927	1	40
714-60-1927	2	40
your SSN	1	60
your SSN	2	55

Figure B-9 Data for HOURS WORKED table

Notice that salaried employees are always given 40 hours. Non-salaried employees (including you) might work any number of hours. For your record, enter your SSN, 60 hours worked for Week 1, and 55 hours worked for Week 2.

CREATING QUERIES

Since you can already create basic queries, this section teaches you the kinds of advanced queries you will create in the Case Studies.

Using Calculated Fields in Queries

A **calculated field** is an output field that is made from *other* field values. A calculated field is *not* a field in a table; it is created in the query generator. The calculated field does not become part of the table—it is just part of query output. The best way to explain this process is by working through an example.

AT THE KEYBOARD

Suppose that you want to see the SSNs and wage rates of hourly workers, and you want to see what the wage rates would be if all employees were given a 10 percent raise. To do this, show the SSN, the current wage rate, and the higher rate (which should be titled "New Rate" in the output). Figure B-10 shows how to set up the query.

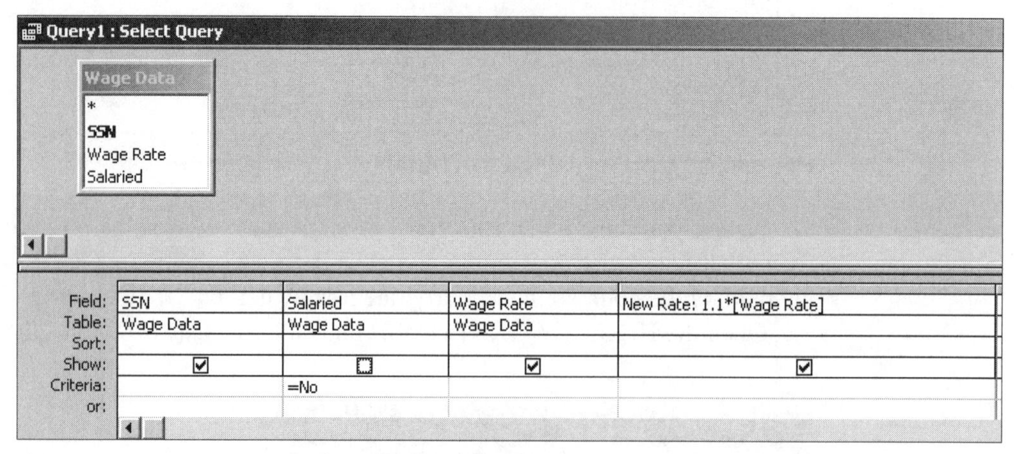

Figure B-10 Query set-up for calculated field

The Salaried field is needed, with the Criteria =No, to select hourly workers. The Show box for that field is checked off, so the Salaried field values will not show in the query output.

Note the expression for the calculated field, which you see in the right-most field cell:

New Rate: 1.1*[Wage Rate]

"New Rate:" merely specifies the desired output heading. (Don't forget the colon.) The 1.1*[Wage Rate] multiplies the old wage rate by 110%, which results in the 10% raise.

In the expression, the field name Wage Rate must be enclosed in square brackets. This is a rule: *Any time that an Access expression refers to a field name, it must be enclosed in square brackets.*

If you run this query, your output should resemble that shown in Figure B-11.

Query1 : Select Query

SSN	Wage Rate	New Rate
114-11-2333	$10.00	11
148-90-1234	$12.00	13.2
099-11-3344	$8.00	8.8

Figure B-11 Output for query with calculated field

Notice that the calculated field output is not shown in Currency format; it's shown as a Double—a number with digits after the decimal point. To convert the output to Currency format, click the line above the calculated field expression, thus activating the column (it darkens). Your data entry screen should resemble the one shown in Figure B-12.

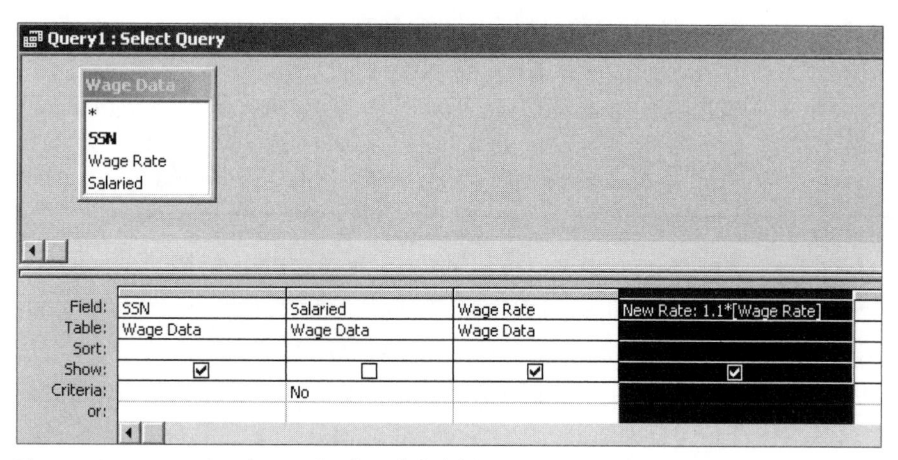

Figure B-12 Activating calculated field in query design

Then select View—Properties. Click the Format drop-down menu. A window, such as the one shown in Figure B-13, will pop up.

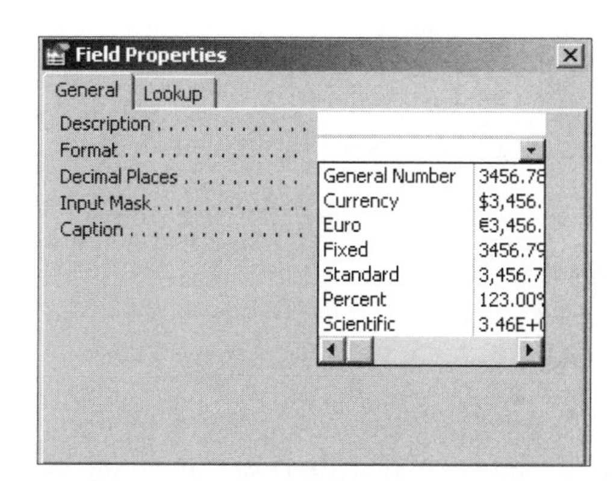

Figure B-13 Field Properties of a calculated field

Click Currency. Then click the upper-right X to close the window. Now when you run the query, the output should resemble that shown in Figure B-14.

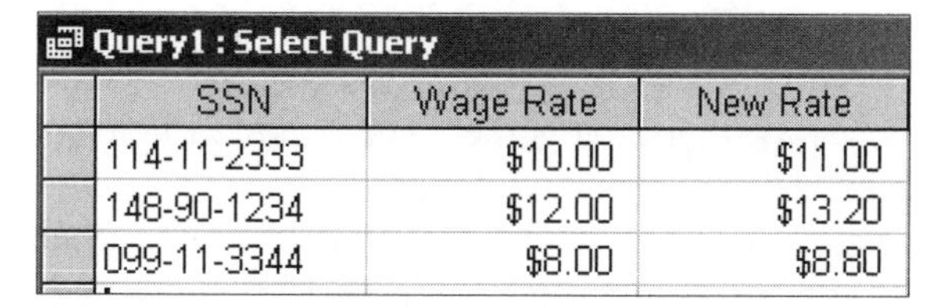

Query1 : Select Query

SSN	Wage Rate	New Rate
114-11-2333	$10.00	$11.00
148-90-1234	$12.00	$13.20
099-11-3344	$8.00	$8.80

Figure B-14 Query output with formatted calculated field

Next, let's look at how to avoid errors when making calculated fields.

Avoiding Errors in Making Calculated Fields

Follow these guidelines to avoid making errors in calculated fields:

- Don't put the expression in the *Criteria* cell, as if the field definition were a filter. You are making a field; so put the expression in the *Field* cell.

- Spell, capitalize, and space a field's name *exactly* as you did in the table definition. If the table definition differs from what you type, Access thinks you're defining a new field by that name. Access then prompts you to enter values for the new field, which it calls a "Parameter Query" field. This is easy to debug because of the tag "Parameter Query." If Access asks you to enter values for a "Parameter," you almost certainly have misspelled a field name in an expression in a calculated field or a criteria.

  Example: Here are some errors you might make for Wage Rate:

  Misspelling: (Wag Rate)

  Case change: (wage Rate / WAGE RATE)

  Spacing change: (WageRate / Wage Rate)

- Don't use parentheses or curly braces instead of the square brackets. Also, don't put parentheses inside square brackets. You *are* allowed to use parentheses outside the square brackets, in the normal algebraic manner.

  Example: Suppose you want to multiply Hours times Wage Rate, to get a field called Wages Owed. This is the correct expression:

  Wages Owed: [Wage Rate]*[Hours]

  This would also be correct:

  Wages Owed: ([Wage Rate]*[Hours])

  But it would **not** be correct to leave out the inside brackets, which is a common error:

  Wages Owed: [Wage Rate*Hours]

"Relating" Two (or More) Tables by the "Join" Operation

Often, the data you need for a query is in more than one table. To complete the query, you must join the tables. One rule of thumb is that joins are made on fields that have common *values,* and those fields can often be key fields. The names of the join fields are irrelevant—the names may be the same, but that is not a requirement for an effective join.

Make a join by first bringing in (Adding) the tables needed. Next, decide which fields you will join. Then, click one field name and hold down the button, dragging the cursor over to the other field's name in its window. Release the button. Access puts a line in, signifying the join. (*Note*: If there are two fields in the tables with the same name, Access will put in the line automatically, so you do not have to do the click-and-drag operation.)

You can join more than two tables together. The common fields *need not* be the same in all tables; that is, you can "daisy chain" them together.

A common join error is to Add a table to the query and then fail to link it to another table. You have a table just "floating" in the top part of the QBE screen! When you run the query, your output will show the same records over and over. This error is unmistakable because there is *so much* redundant output. The rules are (1) add only the tables you need, and (2) link all tables.

Next, you'll work through an example of a query needing a join.

At the Keyboard

Suppose that you want to see the last names, SSNs, wage rates, salary status, and citizenship only for U.S. citizens and hourly workers. The data is spread across two tables, EMPLOYEE and WAGE DATA, so both tables are added, and five fields are pulled down. Criteria are then added. Set up your work to resemble that shown in Figure B-15.

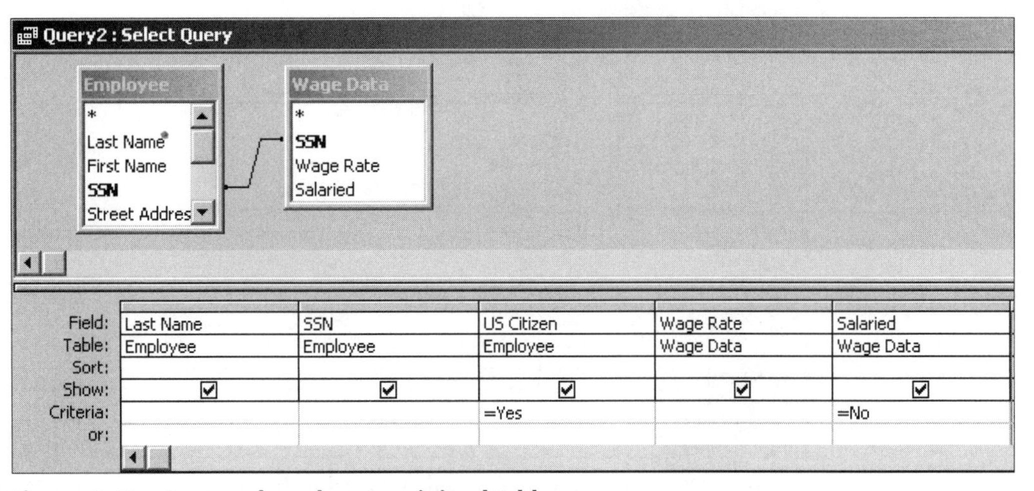

Figure B-15 A query based on two joined tables

In Figure B-15, the join is on the SSN field. A field by that name is in both tables, so Access automatically puts in the join. If one field had been spelled SSN and the other Social Security Number, you would still join on these fields (because of the common values). You would click and drag to do this operation.

Now run the query. The output should resemble that shown in Figure B-16, with the exception of the name Joe Brady.

Query2 : Select Query

Last Name	SSN	US Citizen	Wage Rate	Salaried
Howard	114-11-2333	☑	$10.00	☐
Smith	148-90-1234	☑	$12.00	☐
Brady	099-11-3344	☑	$8.00	☐

Figure B-16 Output of a query based on two joined tables

Here is a quick review of Criteria: If you want data for employees who are U.S. citizens *and* who are hourly workers, the Criteria expressions go into the *same* Criteria row. If you want data for employees who are U.S. citizens *or* who are hourly workers, one of the expressions goes into the second Criteria row (the one that has the "Or:" notation in it).

There is no need to print the query output or to save it. Go back to the Design View and close the query. Another practice query follows.

AT THE KEYBOARD

Suppose that you want to see the wages owed to hourly employees for Week 2. Show the last name, the SSN, the salaried status, week #, and the wages owed. Wages will have to be a calculated field ([Wage Rate]*[Hours]). The criteria are =No for Salaried and =2 for the Week #. (Another "And" query!) You'd set it up the way it is displayed in Figure B-17.

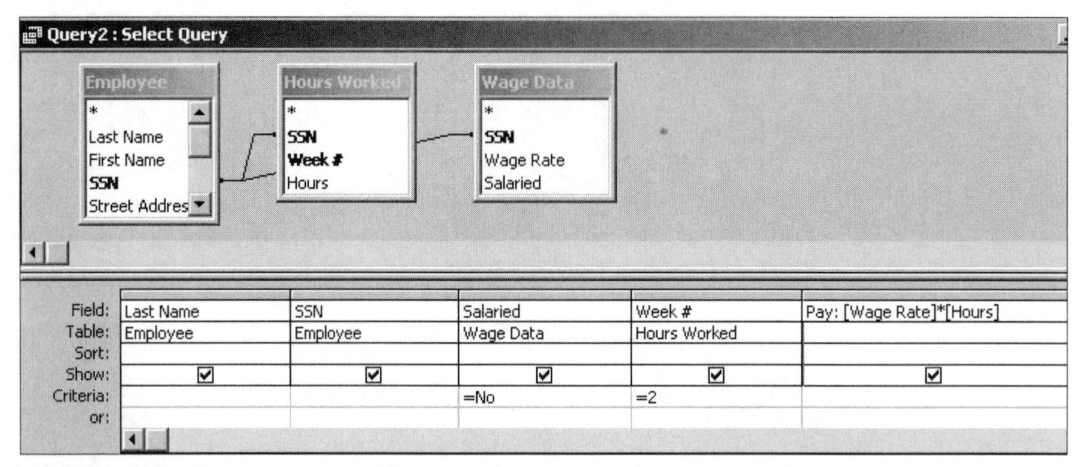

Figure B-17 Query set-up for wages owed to hourly employees for Week 2

In the previous table, the calculated field column was widened so you can see the whole expression. To widen a column, remember to click on the column boundary line and drag to the right.

Run the query. The output should be similar to that shown in Figure B-18 (if you formatted your calculated field to currency).

Query2 : Select Query

Last Name	SSN	Salaried	Week #	Pay
Howard	114-11-2333	☐	2	$500.00
Smith	148-90-1234	☐	2	$480.00
Brady	099-11-3344	☐	2	$440.00

Figure B-18 Query output for wages owed to hourly employees for Week 2

Notice that it was not necessary to pull down the Wage Rate and Hours fields to make this query work. Return to the Design View. There is no need to save. Select File—Close.

Summarizing Data from Multiple Records (Sigma Queries)

You may want data that summarizes values from a field for several records (or possibly all records) in a table. For example, you might want to know the average hours worked for all employees in a week, or perhaps the total (sum of) all the hours worked. Furthermore, you might want data grouped ("stratified") in some way. For example, you might want to know the average hours worked, grouped by all U.S. citizens versus all non-U.S. citizens. Access calls this kind of query a "summary" query, or a **Sigma query**. Unfortunately, this terminology is not intuitive, but the statistical operations that are allowed will be familiar. These operations include the following:

Sum	The total of some field's values
Count	A count of the number of instances in a field, i.e., the number of records. Here, to get the number of employees, you'd count the number of SSN numbers.
Average	The average of some field's values

Min	The minimum of some field's values
Var	The variance of some field's values
StDev	The standard deviation of some field's values

At the keyboard

Suppose that you want to know how many employees are represented in a database. The first step is to bring the EMPLOYEE table into the QBE screen. Do that now. The query will Count the number of SSNs, which is a Sigma query operation. Thus, you must bring down the SSN field.

To tell Access you want a Sigma query, click the little "Sigma" icon in the menu, as shown in Figure B-19.

Figure B-19 Sigma icon

This opens up a new row in the lower part of the QBE screen, called the Total row. At this point, the screen would resemble that shown in Figure B-20.

Figure B-20 Sigma query set-up

Note that the Total cell contain the words "Group By." Until you specify a statistical operation, Access just assumes that a field will be used for grouping (stratifying) data.

To Count the number of SSNs, click next to Group By, revealing a little arrow. Click the arrow to reveal a drop-down menu, as shown in Figure B-21.

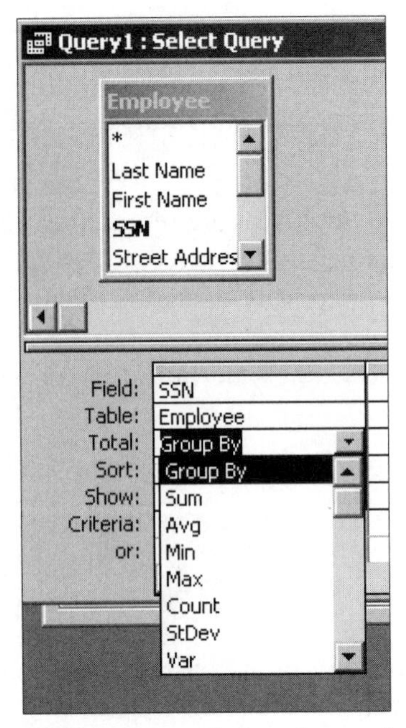

Figure B-21 Choices for statistical operation in a Sigma query

Select the Count operator. (With this menu, you may need to scroll to see the operator you want.) Your screen should now resemble that shown in Figure B-22.

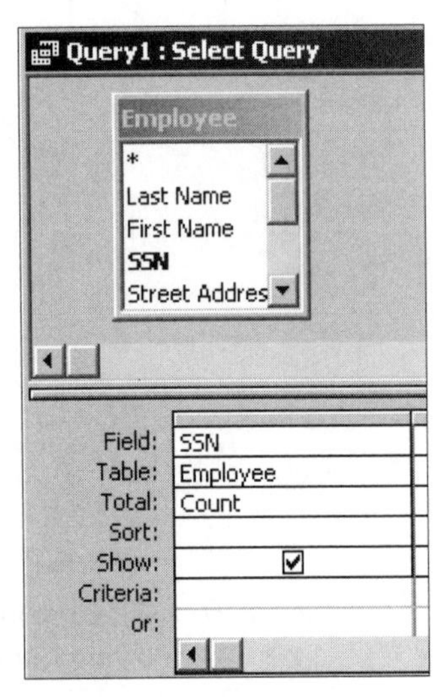

Figure B-22 Count in a Sigma query

Run the query. Your output should resemble that shown in Figure B-23.

Figure B-23 Output of count in a Sigma query

Notice that Access has made a pseudo-heading "CountOfSSN." To do this, Access just spliced together the statistical operation (Count), the word *Of*, and the name of the field

(SSN). What if you wanted an English phrase, such as, "Count of Employees," as a heading? In the Design View, you'd change the query to resemble the one shown in Figure B-24.

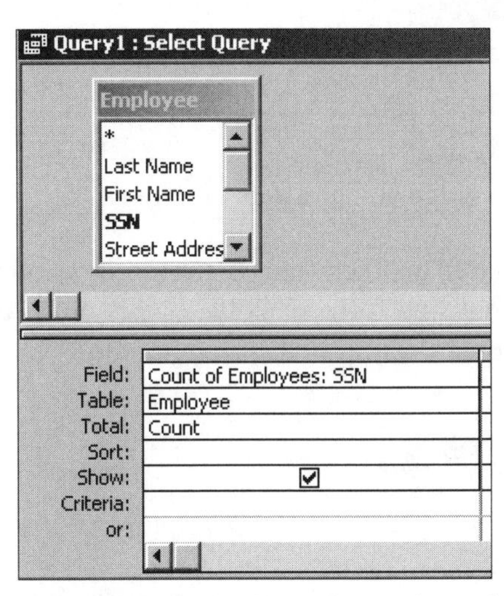

Figure B-24 Heading change in a Sigma query

Now when you run the query, the output should resemble that shown in Figure B-25.

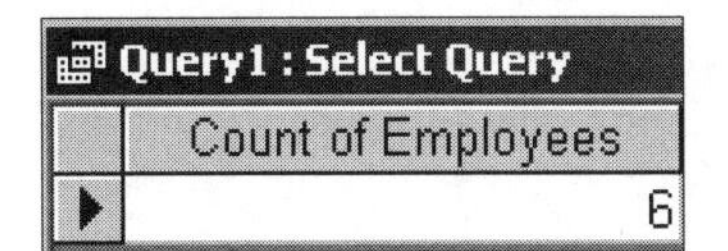

Figure B-25 Output of heading change in a Sigma query

There is no need to save this query. Go back to the Design View and Close.

AT THE KEYBOARD

Here is another example. Suppose that you want to know the average wage rate of employees, grouped by whether they are salaried.

Figure B-26 shows how your query should be set up.

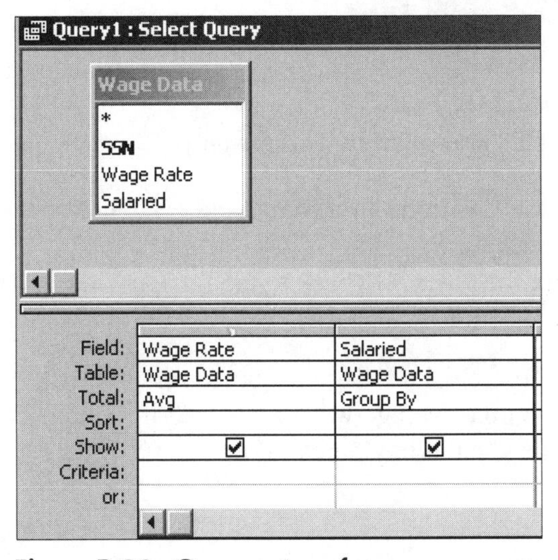

Figure B-26 Query set-up for average wage rate of employees

When you run the query, your output should resemble that shown in Figure B-27.

Query1 : Select Query

	AvgOfWage Rate	Salaried
	$0.00	☑
▶	$10.00	☐

Figure B-27 Output of query for average wage rate of employees

Recall the convention that salaried workers are assigned zero dollars an hour. Suppose that you want to eliminate the output line for zero dollars an hour because only hourly-rate workers matter for this query. The query set-up is shown in Figure B-28.

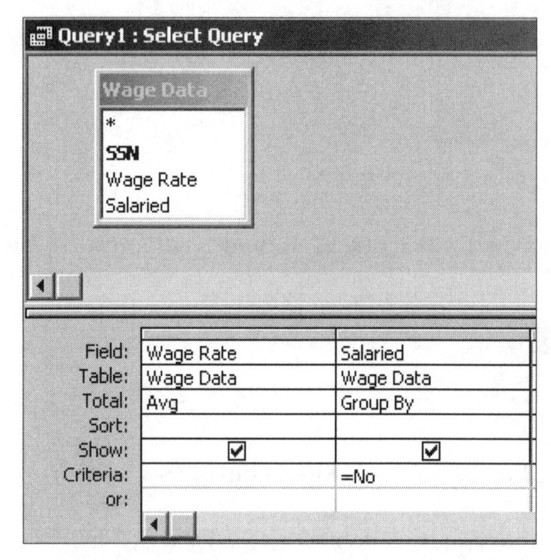

Figure B-28 Query set-up for non-salaried workers only

When you run the query, you'll get output for non-salaried employees only, as shown in Figure B-29.

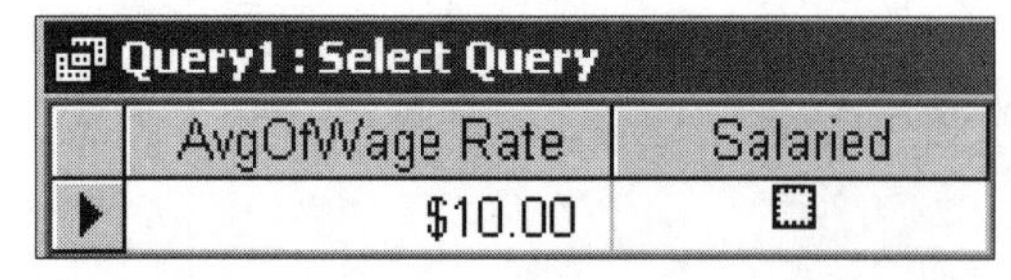

Query1 : Select Query

	AvgOfWage Rate	Salaried
▶	$10.00	☐

Figure B-29 Query output for non-salaried workers only

Thus, it's possible to use a Criteria in a Sigma query without any problem, just as you would with a "regular" query.

There is no need to save the query. Go back to the Design View and Close.

AT THE KEYBOARD

You can make a calculated field in a Sigma query. Assume that you want to see two things for hourly workers: (1) the average wage rate—call it Average Rate in the output; and (2) 110% of this average rate—call it the Increased Rate.

You already know how to do certain things for this query. The revised heading for the average rate would be Average Rate (Average Rate: Wage Rate, in the Field cell). You want the Average of that field. Grouping would be by the Salaried field (with Criteria: =No, for hourly workers).

The most difficult part of this query is to construct the expression for the calculated field. Conceptually it is as follows:

Increased Rate: 1.1*[The current average, however that is denoted]

The question is how to represent [The current average]. You cannot use Wage Rate for this, because that heading denotes the wages before they are averaged. Surprisingly, it turns out that you can use the new heading ("Average Rate") to denote the averaged amount. Thus:

Increased Rate: 1.1*[Average Rate]

Thus, counterintuitively, *you can treat "Average Rate" as if it were an actual field name*. Note, however, that if you use a calculated field, such as Average Rate, in another calculated field, as shown in Figure B-30, you must show that original calculated field in the query output, or the query will ask you to "enter parameter value," which is incorrect. Use the set-up shown in Figure B-30.

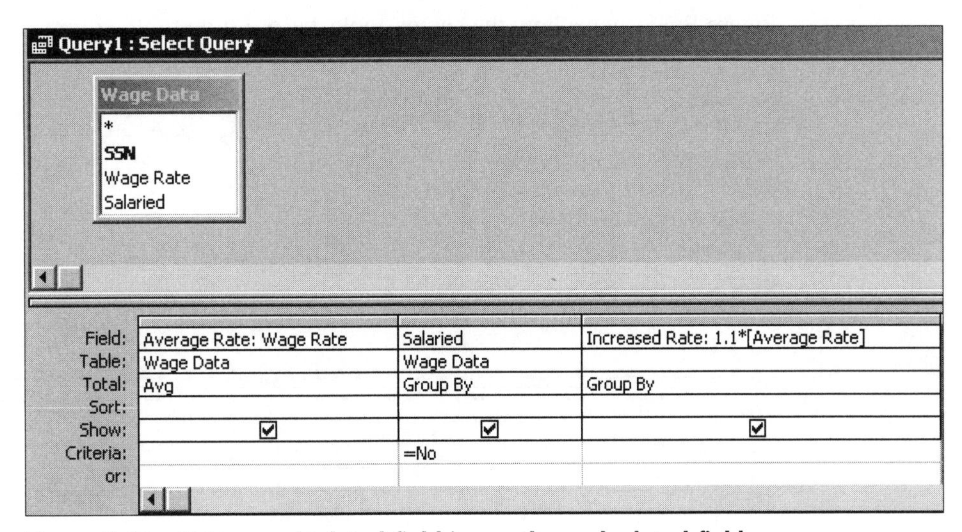

Figure B-30 Using a calculated field in another calculated field

However, if you ran the query now shown in Figure B-30, you'd get some sort of error message. You do not want Group By in the calculated field's Total cell. There is not a *statistical* operator that applies to the calculated field. You must change the Group By operator to Expression. You may have to scroll to get to Expression in the list. Figure B-31 shows how your screen should look.

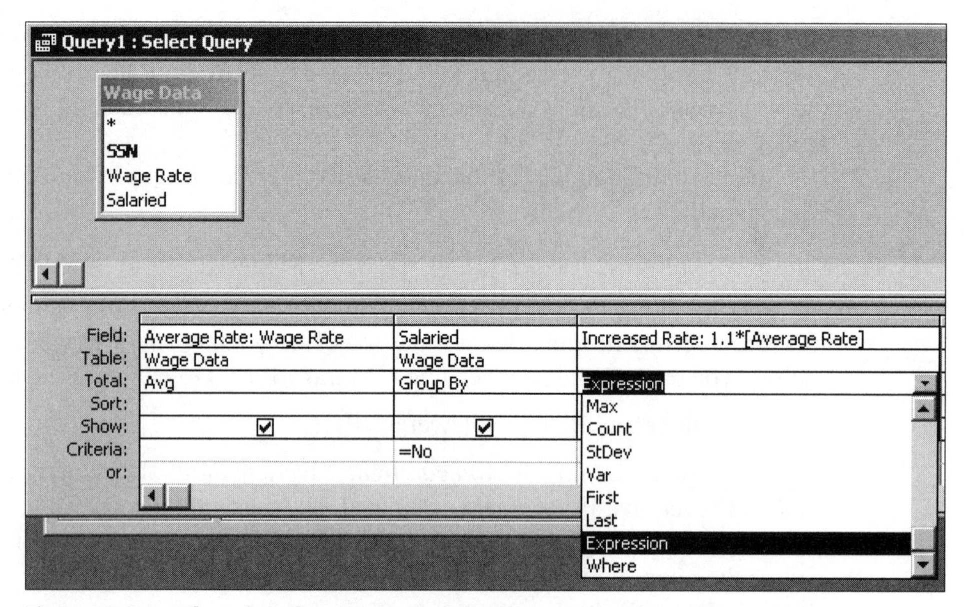

Figure B-31 Changing the Group By to an Expression in a Sigma query

Figure B-32 shows how the screen looks before running the query.

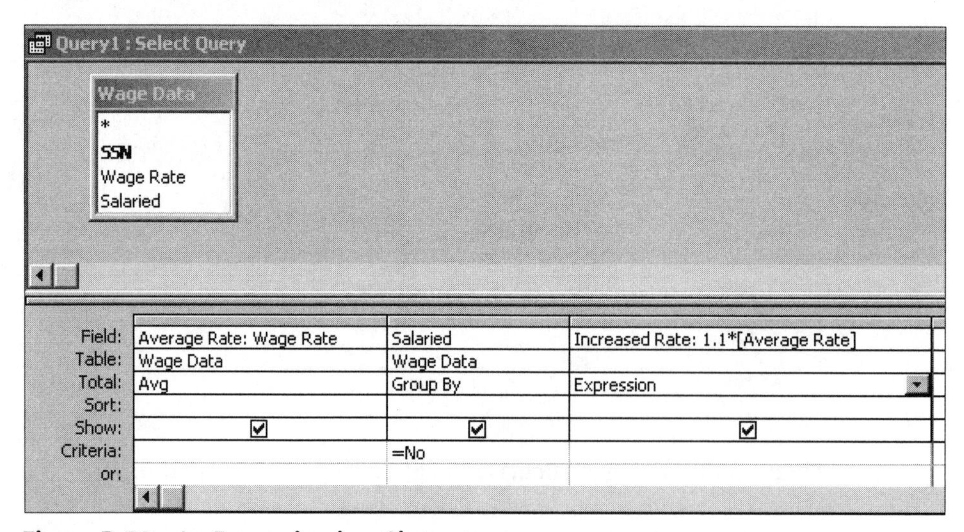

Figure B-32 An Expression in a Sigma query

Figure B-33 shows the output of the query.

Figure B-33 Output of an Expression in a Sigma query

There is no need to save the query definition. Go back to the Design View. Select File—Close.

Using the Date() Function in Queries

Access has two date function features that you should know about. A description of them follows.

1. The following built-in function gives you *today's date*:

   Date()

   You can use this function in a query criteria or in a calculated field. The function "returns" the day on which the query is run. (i.e., it puts that value into the place where the function is in an expression.)

2. *Date arithmetic* lets you subtract one date from another to obtain the number of days difference. Access would evaluate the following expression as the integer 5 (9 less 4 is 5).

   10/9/2004 – 10/4/2004

Here is an example of how date arithmetic works. Suppose that you want to give each employee a bonus equaling a dollar for each day the employee has worked for you. You'd need to calculate the number of days between the employee's date of hire and the day that the query is run, then multiply that number by 1.

The number of elapsed days is shown by the following equation:

Date() – [Date Hired]

Suppose that, for each employee, you want to see the last name, SSN, and bonus amount. You'd set up the query as shown in Figure B-34.

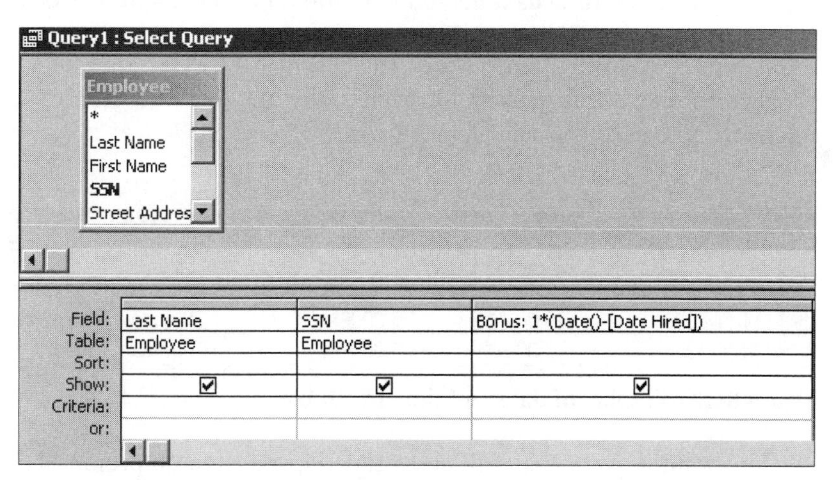

Figure B-34 Date arithmetic in a query

Assume that you set the format of the Bonus field to Currency. The output will be similar to Figure B-35. (Your Bonus data will be different because you are working on a date different than the date when this tutorial was written.)

Query1 : Select Query

Last Name	SSN	Bonus
Brady	099-11-3344	$0.00
Howard	114-11-2333	$144.00
Smith	123-45-6789	$2,396.00
Smith	148-90-1234	$5,640.00
Jones	222-82-1122	$526.00
Ruth	714-60-1927	$1,226.00

Figure B-35 Output of query with date arithmetic

Using Time Arithmetic in Queries

Access will also let you subtract the values of time fields to get an elapsed time. Assume that your database has a JOB ASSIGNMENTS table showing the times that non-salaried employees were at work during a day. The definition is shown in Figure B-36.

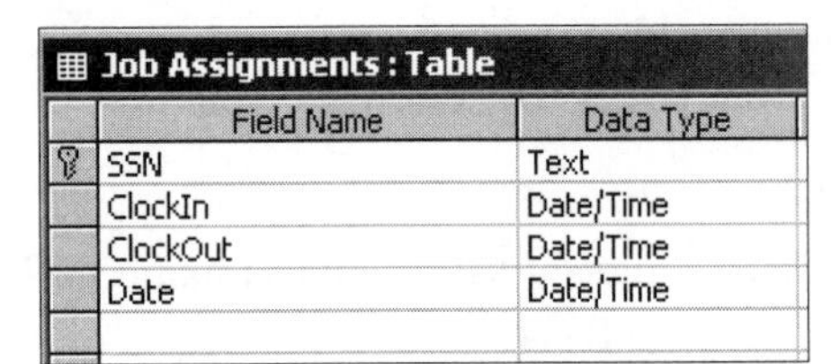

Job Assignments : Table

Field Name	Data Type
SSN	Text
ClockIn	Date/Time
ClockOut	Date/Time
Date	Date/Time

Figure B-36 Date/Time data definition in the JOB ASSIGNMENTS table

Assume that the Date field is formatted for Long Date and that the ClockIn and ClockOut fields are formatted for Medium Time. Assume that, for a particular day, non-salaried workers were scheduled as shown in Figure B-37.

Job Assignments : Table

SSN	ClockIn	ClockOut	Date
099-11-3344	8:30 AM	4:30 PM	Thursday, September 30, 2004
114-11-2333	9:00 AM	3:00 PM	Thursday, September 30, 2004
148-90-1234	7:00 AM	5:00 PM	Thursday, September 30, 2004

Figure B-37 Display of date and time in a table

You want a query that will show the elapsed time on premises for the day. When you add the tables, your screen may show the links differently. Click and drag the JOB ASSIGNMENTS, EMPLOYEE, and WAGE DATA table icons to look like those in Figure B-38.

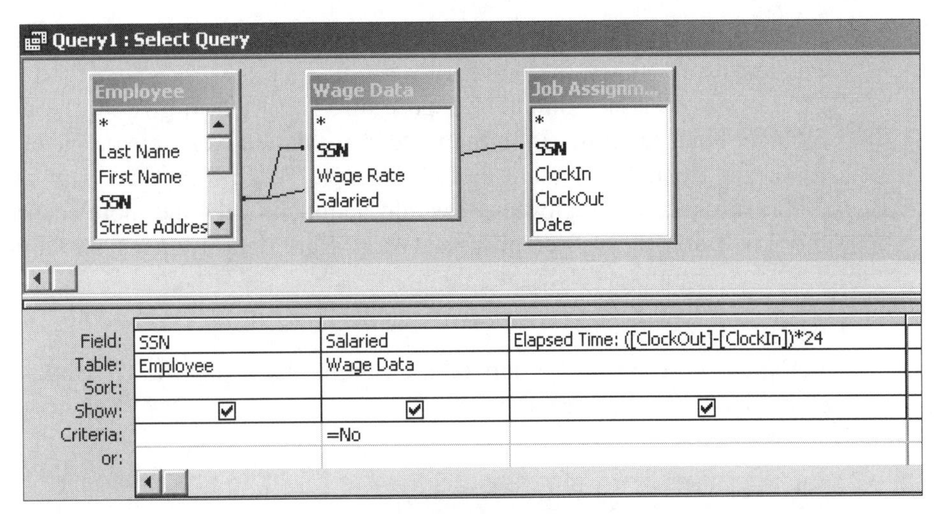

Figure B-38 Query set-up for time arithmetic

Figure B-39 shows the output.

Query1 : Select Query

SSN	Salaried	Elapsed Time
114-11-2333	☐	6
148-90-1234	☐	10
099-11-3344	☐	8

Figure B-39 Query output for time arithmetic

The output looks right. For example, employee 099-11-3344 was at work from 8:30 a.m. to 4:30 p.m., which is eight hours. But how does the odd expression that follows yield the correct answers?

([ClockOut] – [ClockIn]) * 24

Why wouldn't the following expression, alone, work?

[ClockOut] – [ClockIn]

This is the answer: In Access, *subtracting one time from the other yields the decimal portion of a 24-hour day*. Employee 099-11-3344 worked 8 hours, which is a third of a day, so .3333 would result. That is why you must multiply by 24—to convert to an hour basis. Continuing with 099-11-3344, 1/3 x 24 = 8.

Note that parentheses are needed to force Access to do the subtraction *first*, before the multiplication. Without parentheses, multiplication takes precedence over subtraction. With the following expression, ClockIn would be multiplied by 24 and then that value would be subtracted from ClockOut, and the output would be a nonsense decimal number:

[ClockOut] – [ClockIn] * 24

Delete and Update Queries

Thus far, the queries presented in this tutorial have been Select queries. They select certain data from specific tables, based on a given criterion. You can also create queries to update the original data in a database. Businesses do this often, and in real time. For example, when you order an item from a Web site, the company's database is updated to reflect the purchase of the item by deleting it from inventory.

Let's look at an example. Suppose that you want to give all the non-salaried workers a $.50 per hour pay raise. With the three non-salaried workers you have now, it would be easy to simply go into the table and change the Wage Rate data. But assume that you have 3,000 non-salaried employees. It would be much faster and more accurate to change each of the 3,000 non-salaried employees' Wage Rate data by using an Update query to add the $.50 to each employee's wage rate.

AT THE KEYBOARD

Let's change each of the non-salaried employees' pay via an Update query. Figure B-40 shows how to set up the query.

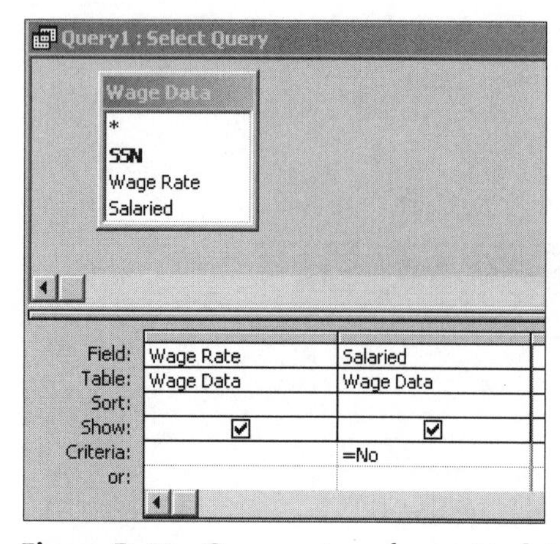

Figure B-40 Query set-up for an Update Query

So far, this query is just a Select query. Place your cursor somewhere above the QBE grid, and then right-click the mouse. Once in that menu, choose Query Type—Update Query, as shown in Figure B-41.

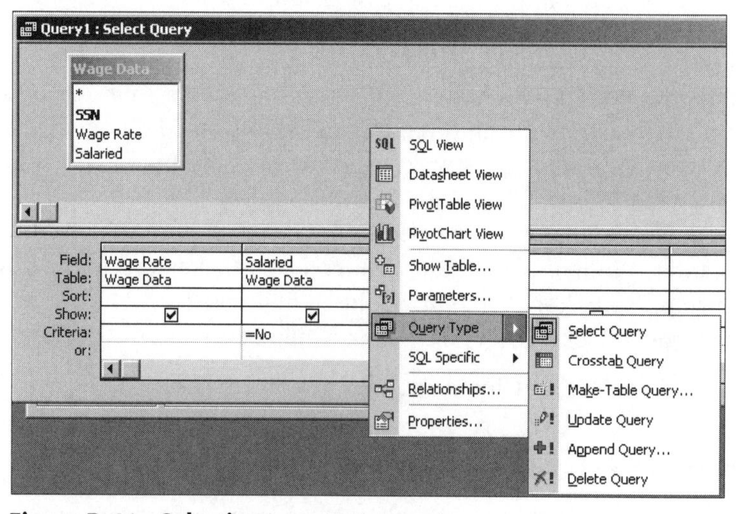

Figure B-41 Selecting a query type

Notice that you now have another line on the QBE grid called "Update to." This is where you specify the change or update to the data. Notice that you are going to update only the non-salaried workers by using a filter under the Salaried field. Update the Wage Rate data to Wage Rate plus $.50, as shown in Figure B-42 . (Note the [ ] as in a calculated field.)

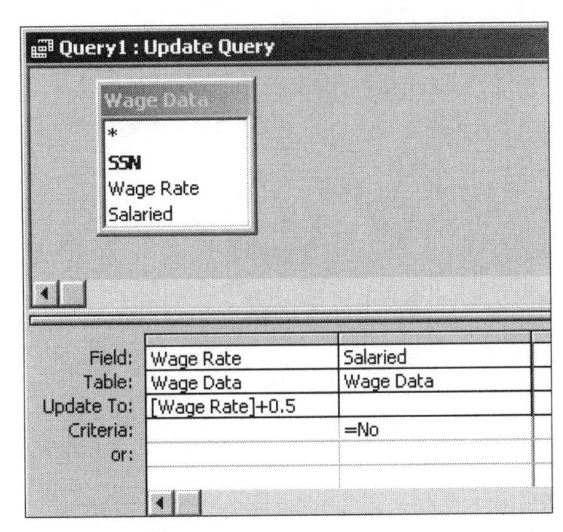

Figure B-42 Updating the wage rate for non-salaried workers

Now run the query. You will first get a warning message, as shown in Figure B-43.

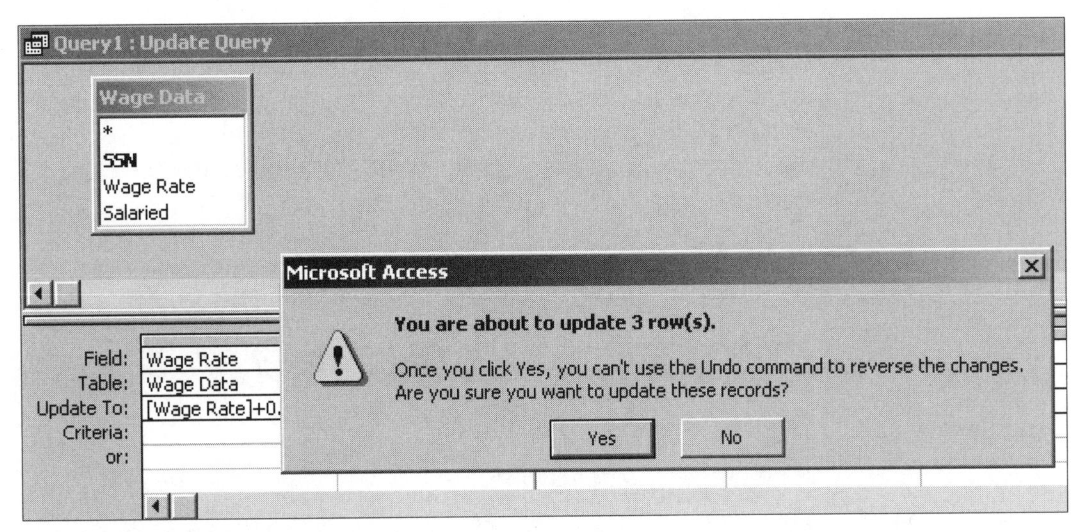

Figure B-43 Update query warning

Once you click "yes," the records will be updated. Check those updated records now by viewing the WAGE DATA table. Each salaried wage rate should now be increased by $.50. Note that in this example, you are simply adding $.50 to each salaried wage rate. You could add or subtract data from another table as well. If you do that, remember to call the field name in square brackets.

Delete queries work the same way as Update queries. Assume that your company has been taken over by the state of Delaware. The state has an odd policy of only employing Delaware residents. Thus, you must delete (or fire) all employees who are not Delaware residents. To do this, you would first create a Select query, using the EMPLOYEE table, right-click your mouse, choose Delete Query from Query Type, then bring down the State field and filter only those records not in Delaware (DE). Do not perform this operation, but if you did, the set-up would look like that in Figure B-44.

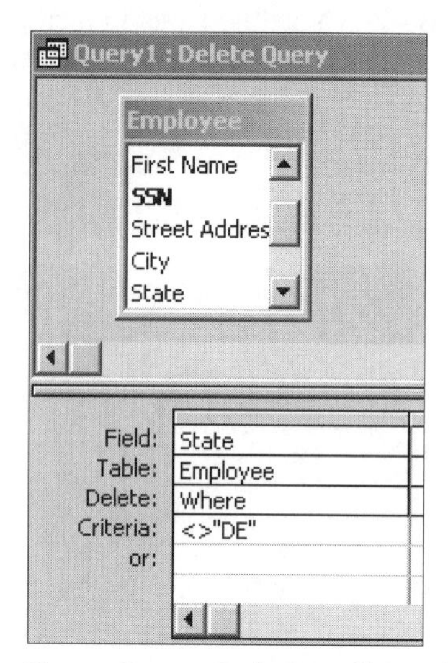

Figure B-44 Deleting all employees who are not Delaware residents

Parameter Queries

Another type of query, which is a type of Select query, is a called a **Parameter query**. Here is an example. Suppose that your company has 5,000 employees. You might want to query the database to find the same kind of information again and again, only about different employees. For example, you might want to query the database to find out how many hours a particular employee has worked. To do this, you could run a query previously created and stored, but run it only for a particular employee.

At the Keyboard

Create a Select query with the format shown in Figure B-45.

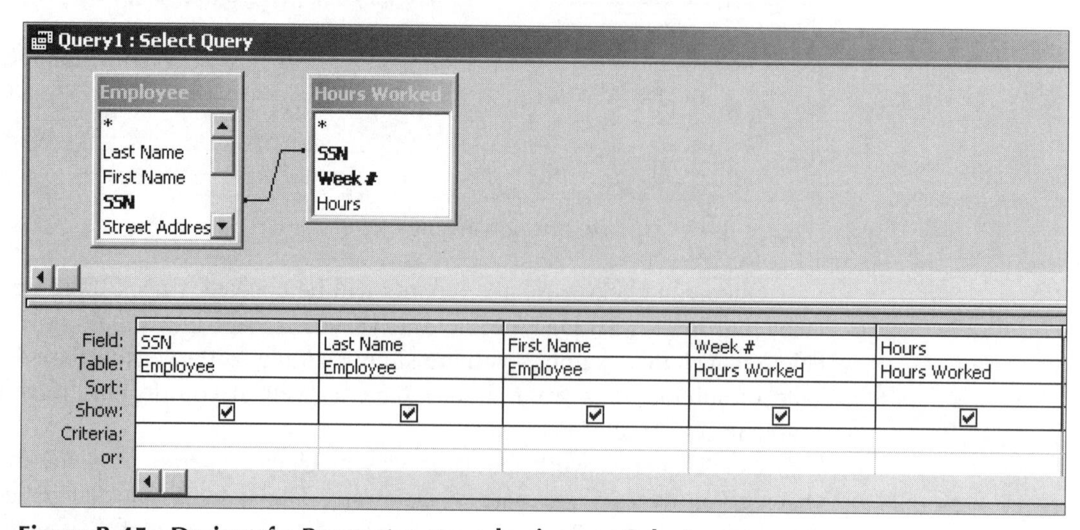

Figure B-45 Design of a Parameter query begins as a Select query

In the criteria line of the QBE grid for the field SSN, type in the following as shown in Figure B-46.

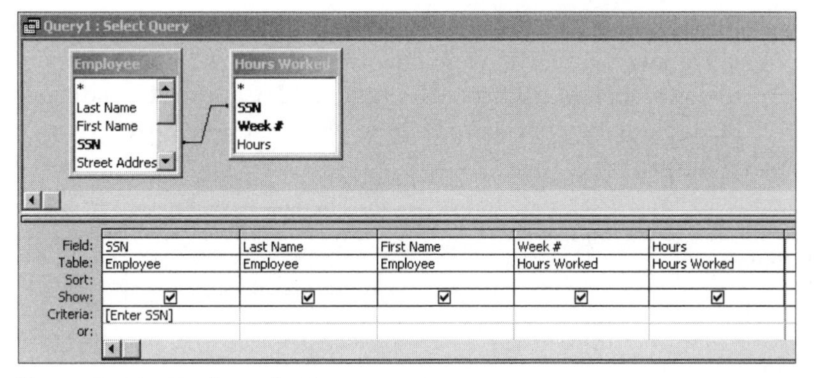

Figure B-46 Design of a Parameter query

Note the square brackets, like you would expect to see in a calculated field.

Now run that query. You will be prompted for the specific employee's SSN as shown in Figure B-47.

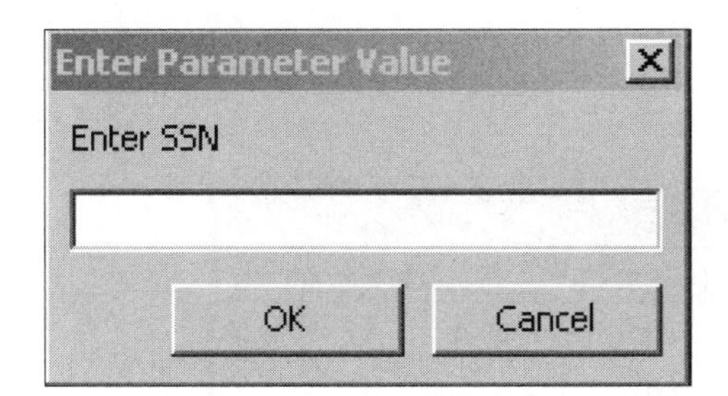

Figure B-47 Enter Parameter Value dialog box

Type in your own SSN. Your query output should resemble that shown in Figure B-48.

Query1 : Select Query

SSN	Last Name	First Name	Week #	Hours
099-11-3344	Brady	Joe	1	60
099-11-3344	Brady	Joe	2	55

Figure B-48 Output of a Parameter query

Seven Practice Queries

This portion of the tutorial is designed to provide you with additional practice in making queries. Before making these queries, you must create the specified tables and enter the records shown in the Creating Tables section of this tutorial. The output shown for the practice queries is based on those inputs.

At the Keyboard

For each query that follows, you are given a problem statement and a "scratch area." You are also shown what the query output should look like. Follow this procedure: Set up a query in Access. Run the query. When you are satisfied with the results, save the query and continue with the next query. You will be working with the EMPLOYEE, HOURS WORKED, and WAGE DATA tables.

1. Create a query that shows the SSN, last name, state, and date hired for those living in Delaware *and* who were hired after 12/31/92. Sort (ascending) by SSN. (Sorting review: Click in the Sort cell of the field. Choose Ascending or Descending.) Use the table shown in Figure B-49 to work out your QBE grid on paper before creating your query.

Field					
Table					
Sort					
Show					
Criteria					
Or:					

Figure B-49 QBE grid template

Your output should resemble that shown in Figure B-50.

Query1 : Select Query

SSN	Last Name	State	Date Hired
114-11-2333	Howard	DE	8/1/2004
123-45-6789	Smith	DE	6/1/1996
222-82-1122	Jones	DE	7/15/2001

Figure B-50 Number 1 query output

2. Create a query that shows the last name, first name, date hired, and state for those living in Delaware *or* who were hired after 12/31/92. The primary sort (ascending) is on last name, and secondary sort (ascending) is on first name. (Review: The Primary Sort field must be left of the Secondary Sort field in the query set-up.) Use the table shown in Figure B-51 to work out your QBE grid on paper before creating your query.

Field					
Table					
Sort					
Show					
Criteria					
Or:					

Figure B-51 QBE grid template

If your name were Joseph Brady, your output would look like that shown in Figure B-52.

Last Name	First Name	Date Hired	State
Brady	Joe	12/23/2002	MN
Howard	Jane	8/1/2004	DE
Jones	Sue	7/15/2001	DE
Ruth	Billy	8/15/1999	MD
Smith	Albert	7/15/1987	DE
Smith	John	6/1/1996	DE

Figure B-52 Number 2 query output

3. Create a query that shows the sum of hours worked by U.S. citizens and by non-U.S. citizens (i.e., group on citizenship). The heading for total hours worked should be Total Hours Worked. Use the table shown in Figure B-53 to work out your QBE grid on paper before creating your query.

Field					
Table					
Total					
Sort					
Show					
Criteria					
Or:					

Figure B-53 QBE grid template

Your output should resemble that shown in Figure B-54.

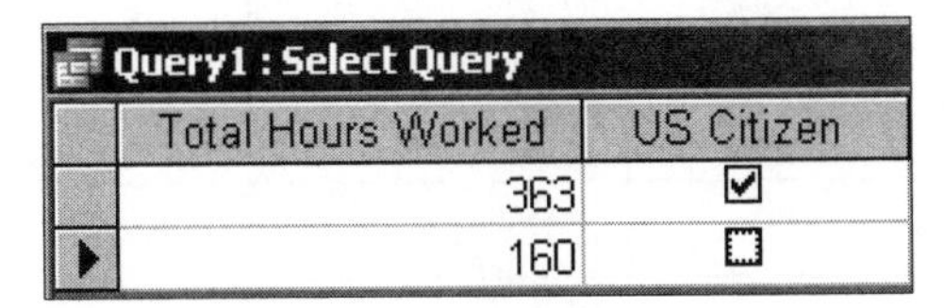

Total Hours Worked	US Citizen
363	☑
160	☐

Figure B-54 Number 3 query output

4. Create a query that shows the wages owed to hourly workers for week 1. The heading for the wages owed should be Total Owed. The output headings should be: Last Name, SSN, Week #, and Total Owed. Use the table shown in Figure B-55 to work out your QBE grid on paper before creating your query.

Field					
Table					
Sort					
Show					
Criteria					
Or:					

Figure B-55 QBE grid template

If your name were Joseph Brady, your output would look like that in Figure B-56.

Query1 : Select Query

Last Name	SSN	Week #	Total Owed
Howard	114-11-2333	1	$420.00
Smith	148-90-1234	1	$475.00
Brady	099-11-3344	1	$510.00

Figure B-56 Number 4 query output

5. Create a query that shows the last name, SSN, hours worked, and overtime amount owed for employees paid hourly who earned overtime during Week 2. Overtime is paid at 1.5 times the normal hourly rate for hours over 40. The amount shown should be just the overtime portion of the wages paid. This is not a Sigma query—amounts should be shown for individual workers. Use the table shown in Figure B-57 to work out your QBE grid on paper before creating your query.

Field					
Table					
Sort					
Show					
Criteria					
Or:					

Figure B-57 QBE grid template

If your name were Joseph Brady, your output would look like that shown in Figure B-58.

Query1 : Select Query

Last Name	SSN	Hours	OT Pay
Howard	114-11-2333	50	$157.50
Brady	099-11-3344	55	$191.25

Figure B-58 Number 5 query output

6. Create a parameter query that shows the hours employees have worked. Have the parameter query prompt for the week number. The output headings should be Last Name, First Name, Week #, and Hours. Do this only for the non-salaried workers. Use the table shown in Figure B-59 to work out your QBE grid on paper before creating your query.

Field					
Table					
Sort					
Show					
Criteria					
Or:					

Figure B-59 QBE grid template

Run the query with "2" when prompted for the Week #. Your output should look like that shown in Figure B-60.

Query1 : Select Query

Last Name	First Name	Week #	Hours
Howard	Jane	2	50
Smith	Albert	2	40
Brady	Joe	2	55

Figure B-60 Number 6 query output

7. Create an update query that gives certain workers a merit raise. You must first create an additional table as shown in Figure B-61.

Merit Raises : Table

SSN	Merit Raise
114-11-2333	$0.25
148-90-1234	$0.15

Figure B-61 MERIT RAISES table

Now make a query that adds the Merit Raise to the current Wage Rate for those who will receive a raise. When you run the query, you should be prompted with "You are about to update two rows." Check the original WAGE DATA table to confirm the update. Use the table shown in Figure B-62 to work out your QBE grid on paper before creating your query.

Field					
Table					
Update to					
Criteria					
Or:					

Figure B-62 QBE grid template

CREATING REPORTS

Database packages let you make attractive management reports from a table's records or from a query's output. If you are making a report from a table, the Access report generator looks up the data in the table and puts it into report format. If you are making a report from a query's output, Access runs the query in the background (you do not control this or see this happen) and then puts the output in report format.

There are three ways to make a report. One is to handcraft the report in the so-called "Design View," from scratch. This is tedious and is not shown in this tutorial. The second way is to use the so-called "Report Wizard," during which Access leads you through a menu-driven construction. This method is shown in this tutorial. The third way is to start in the Wizard and then use the Design View to tailor what the Wizard produces. This method is also shown in this tutorial.

Creating a Grouped Report

This tutorial assumes that you can use the Wizard to make a basic ungrouped report. This section of the tutorial teaches you how to make a grouped report. (If you cannot make an ungrouped report, you might learn how to make one by following the first example that follows.)

AT THE KEYBOARD

Suppose that you want to make a report out of the HOURS WORKED table. At the main objects menu, start a new report by choosing Reports—New. Select the Report Wizard and select the HOURS WORKED table from the drop-down menu as the report basis. Select OK. In the next screen, select all the fields (using the >> button), as shown in Figure B-63.

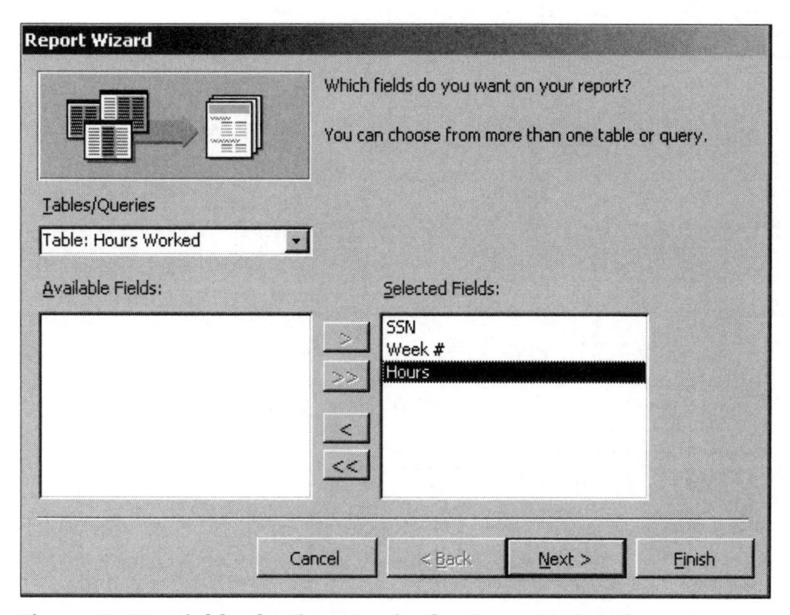

Figure B-63 Field selection step in the Report Wizard

Click Next. Then tell Access that you want to group on Week # by double-clicking that field name. You'll see that shown in Figure B-64.

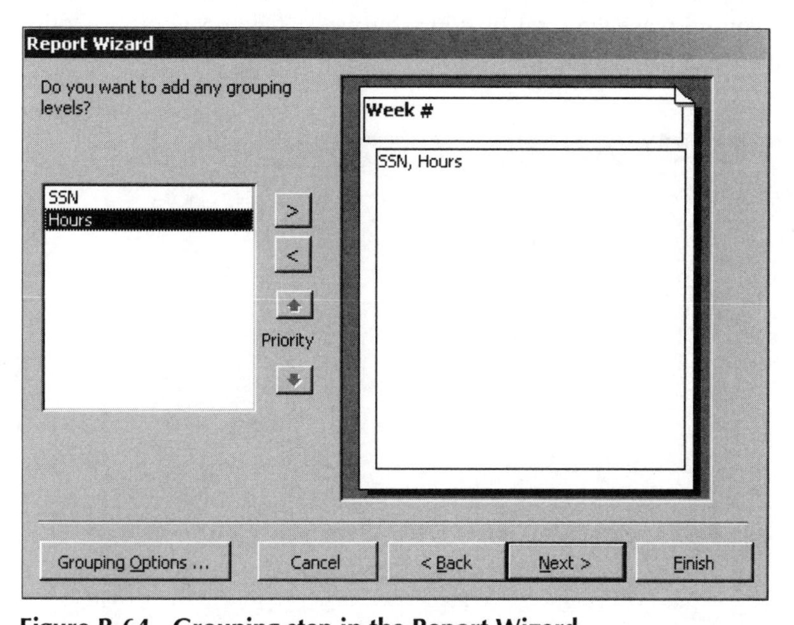

Figure B-64 Grouping step in the Report Wizard

Click Next. You'll see a screen, similar to the one shown in Figure B-65, for Sorting and for Summary Options.

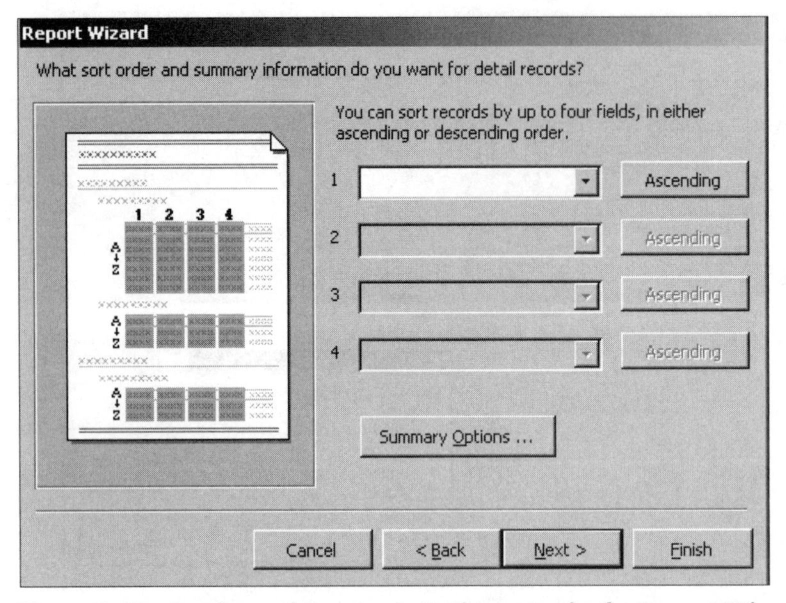

Figure B-65 Sorting and Summary Options step in the Report Wizard

Because you chose a grouping field, Access will now let you decide whether you want to see group subtotals and/or report grand totals. All numeric fields could be added, if you choose that option. In this example, group subtotals are for total hours in each week. Assume that you *do* want the total of hours by week. Click Summary Options. You'll get a screen similar to the one in Figure B-66.

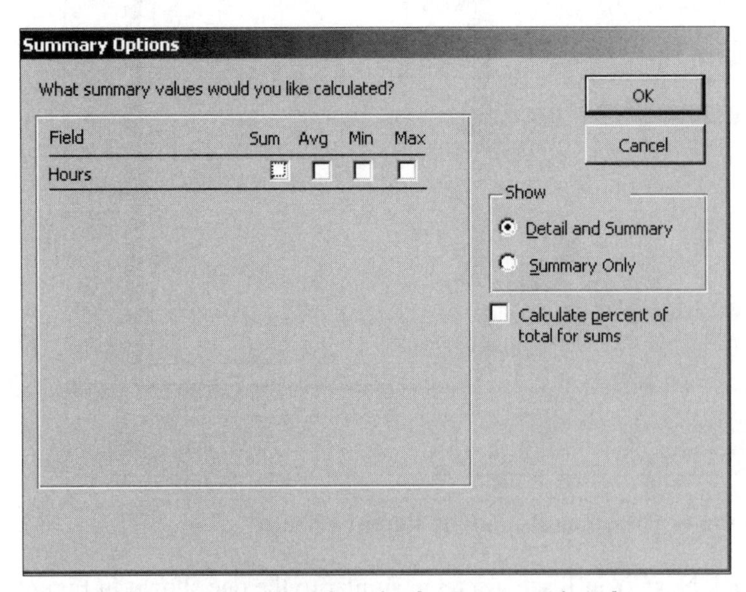

Figure B-66 Summary Options in the Report Wizard

Next, follow these steps:

1. Click the Sum box for Hours (to sum the hours in the group).
2. Click Detail and Summary. (Detail equates with "group," and Summary with "grand total for the report.")
3. Click OK. This takes you back to the Sorting screen, where you can choose an ordering within the group, if desired. (In this case, none is.)

4. Click Next to continue.
5. In the Layout screen (not shown here) choose Stepped and Portrait.
6. Click *off* the "Fit on a page" option.
7. Click Next.
8. In the Style screen (not shown), accept Corporate.
9. Click Next.
10. Provide a title—Hours Worked by Week would be appropriate.
11. Select the Preview button to view the report.
12. Click Finish.

The top portion of your report will look like that shown in Figure B-67.

Hours Worked by Week

Week #	*SSN*	*Hours*
1		
	099-11-3344	60
	714-60-1927	40
	222-82-1122	40
	148-90-1234	38
	123-45-6789	40
	114-11-2333	40
Summary for 'Week #' = 1 (6 detail records)		
Sum		*258*

Figure B-67 Hours Worked by Week report

Notice that data is shown grouped by weeks, with Week 1 on top, then a subtotal for that week. Week 2 data is next, then there is a grand total (which you can scroll down to see). The subtotal is labeled "Sum," which is not very descriptive. This can be changed later in the Design View. Also, there is the apparently useless italicized line that starts out *"Summary for 'Week ..."* This also can be deleted later in the Design View. At this point, you should select File—SaveAs (accept the suggested title if you like). Then select File—Close to get back the main objects menu. Try it. Your report's objects screen should resemble that shown in Figure B-68.

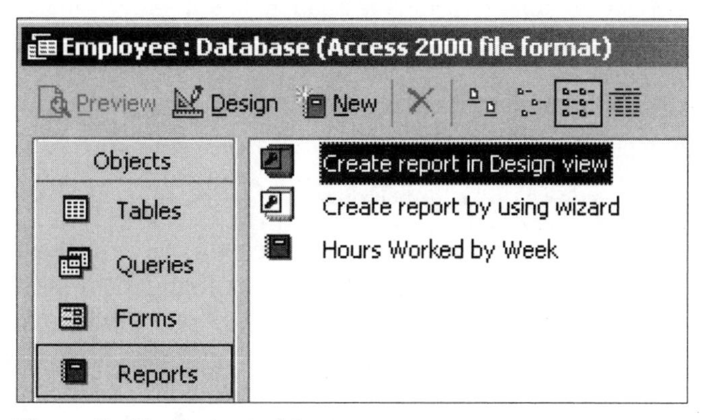

Figure B-68 Report objects screen

To edit the report in the Design View, click the report title, then the Design button. You will see a complex (and intimidating) screen, similar to the one shown in Figure B-69.

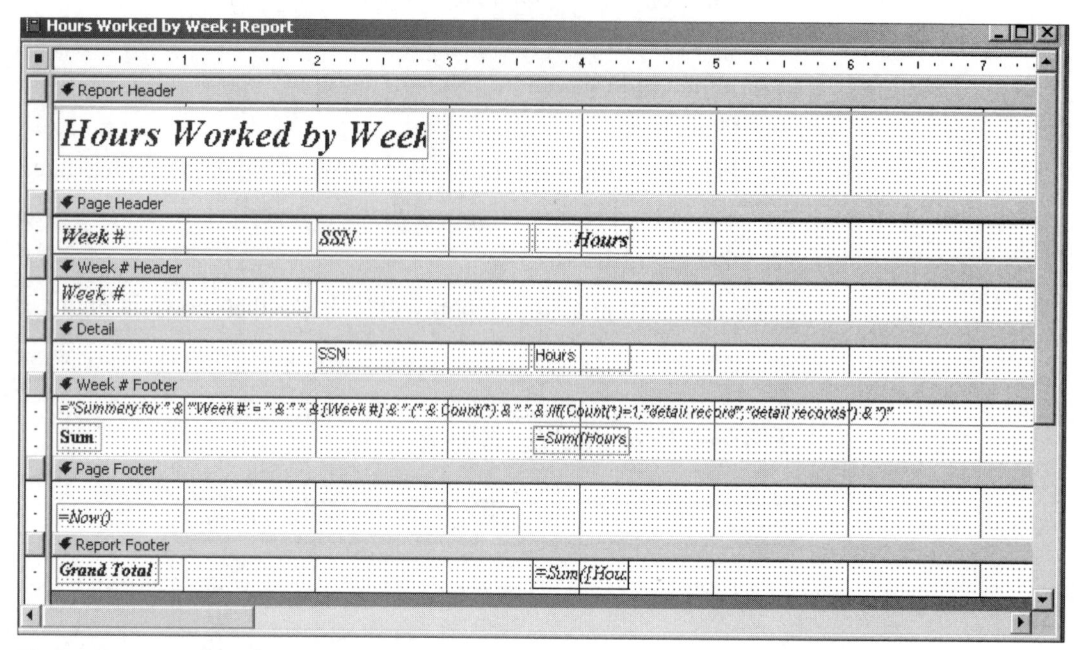

Figure B-69 Report design screen

The organization of the screen is hierarchical. At the top is the Report level. The next level down (within a report) is the Page level. The next level or levels down (within a page) are for any data groupings you have specified.

If you told Access to make group (summary) totals, your report will have a "header" and a grand report total. The report header is usually just the title you have specified, and often the date is put in by default.

A page also has a header, which is usually just the names of the fields you have told Access to put in the report (here, Week #, SSN, and Hours fields). Sometimes the page number is put in by default.

Groupings of data are more complex. There is a header for the group—in this case, the *value* of the Week # will be the header; for example, there is a group of data for the first week, then one for the second—the values shown will be 1 and 2. Within each data grouping is the other "detail" that you've requested. In this case, there will be data for each SSN and the related hours.

Each Week # gets a "footer," which is a labeled sum—recall that you asked for that to be shown (Detail and Summary were requested). The Week # footer is indicated by three things:

1. The italicized line that starts "=Summary for ..."
2. The Sum label
3. The adjacent expression "=Sum(Hours)"

The italicized line beneath the Week # footer will be printed unless you eliminate it. Similarly, the word "Sum" will be printed as the subtotal label unless you eliminate it. The "=Sum(Hours)" is an expression that tells Access to add up the quantity *for the header in question* and put that number into the report as the subtotal. (In this example, that would be the sum of hours, by Week #.)

Each report also gets a footer—the grand total (in this case, of hours) for the report.

If you look closely, each of the detail items appears to be doubly inserted in the design. For example, you will see the notation for SSN twice, once in the Page Header and then again in the Detail band. Hours are treated similarly.

The data items will not actually be printed twice, because each data element is an object in the report; each object is denoted by a label and by its value. There is a representation of the name, which is the boldface name itself (in this example, "SSN" in the page header), and there is a representation in less-bold type for the value "SSN" in the Detail band.

Sometimes, the Report Wizard is arbitrary about where it puts labels and data. However, if you do not like where the Wizard puts data, it can be moved around in the Design View. You can click and drag within the band or across bands. Often, a box will be too small to allow full numerical values to show. When that happens, select the box and then click one of the sides to stretch it. This will allow full values to show. At other times an object's box will be very long. When that happens, the box can be clicked, re-sized, then dragged right or left in its panel to reposition the output.

Suppose that you do *not* want the italicized line to appear in the report. Also suppose that you would like different subtotal and grand total labels. The italicized line is an object that can be activated by clicking it. Do that. "Handles" (little squares) appear around its edges, as shown in Figure B-70.

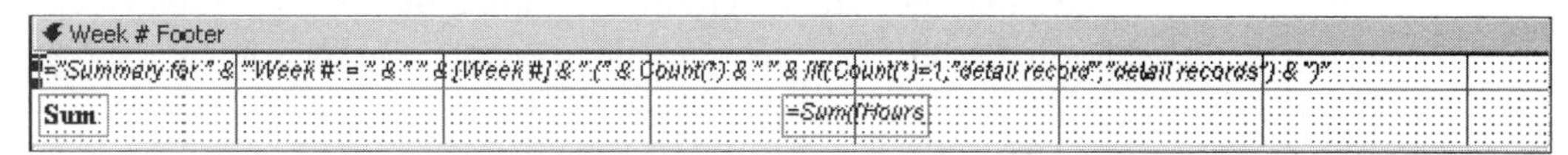

Figure B-70 Selecting an object in the Report Design View

Press the Delete key to get rid of the highlighted object.

To change the subtotal heading, click the Sum object, as shown in Figure B-71.

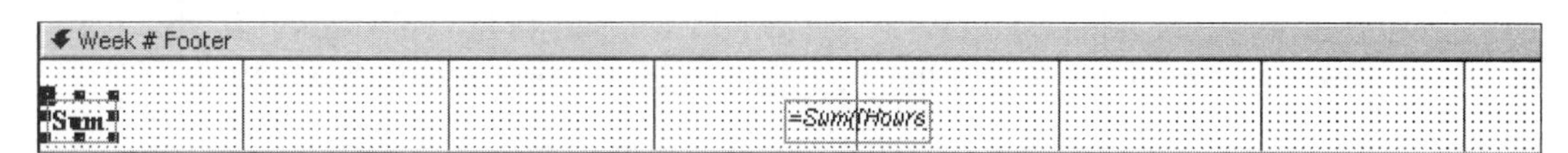

Figure B-71 Selecting the Sum object in the Report Design View

Click again. This gives you an insertion point from which you can type, as shown in Figure B-72.

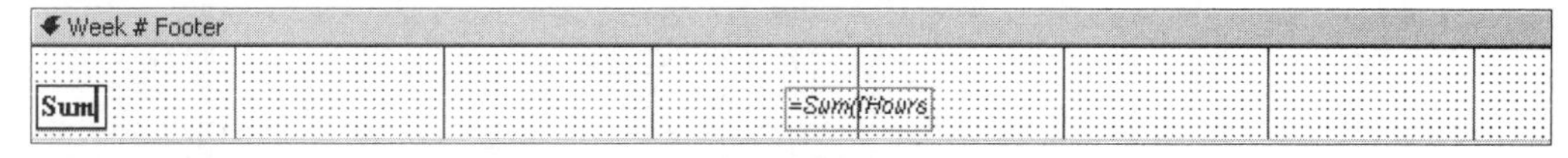

Figure B-72 Typing in an object in the Report Design View

Change the label to something like Sum of Hours for Week, then hit Enter, or click somewhere else in the report to deactivate. Your screen should resemble that shown in Figure B-73.

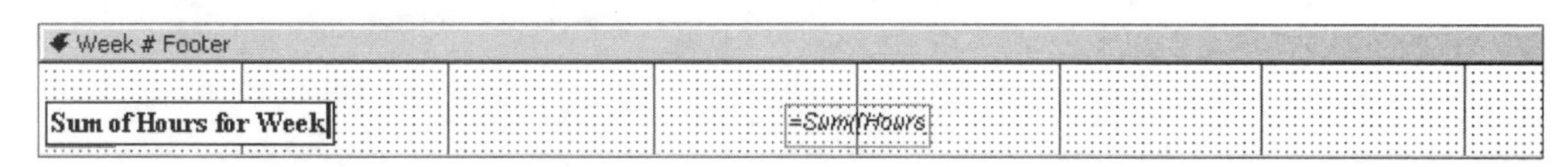

Figure B-73 Changing a label in the Report Design View

You can change the Grand Total in the same way.

Finally, you'll want to save and then print the file: File—Save. Then select File—Print Preview. You should see a report similar to that in Figure B-74 (top part is shown).

Hours Worked by Week

Week #	*SSN*	*Hours*
	1	
	099-11-3344	60
	714-60-1927	40
	222-82-1122	40
	148-90-1234	38
	123-45-6789	40
	114-11-2333	40
Sum of Hours for Week		258

Figure B-74 Hours Worked by Week report

Notice that the data are grouped by week number (data for Week 1 is shown) and subtotaled for that week. The report would also have a grand total at the bottom.

Moving Fields in the Design View

When you group on more than one field in the Report Wizard, the report has an odd "staircase" look. There is a way to overcome that effect in the Design View, which you will learn next.

Suppose that you make a query showing an employee's last name, street address, zip code and wage rate. Then you make a report from that query, grouping on last name, street address, and zip code. (Why you would want to organize a report in this way is not clear, but for the moment, accept the organization for the purpose of the example.) This is shown in Figure B-75.

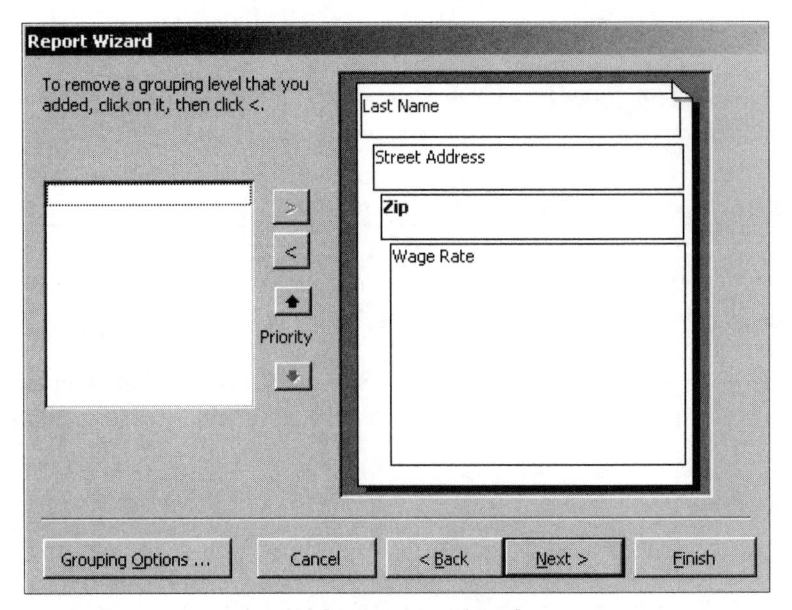

Figure B-75 Grouping in the Report Wizard

Then, follow these steps:

1. Click Next.
2. You do not Sum anything in Summary Options.
3. Click *off* the "Fit on a Page" option.
4. Select Landscape.
5. Select Stepped. Click Next.
6. Select Corporate. Click Next.
7. Type a title (Wage Rates for Employees). Click Finish.

When you run the report, it will have a "staircase" grouped organization. In the report that follows in Figure B-76, notice that Zip data is shown below Street Address data, and Street Address data is shown below Last Name data. (The field Wage Rate is shown subordinate to all others, as desired. Wage rates may not show on the screen without scrolling).

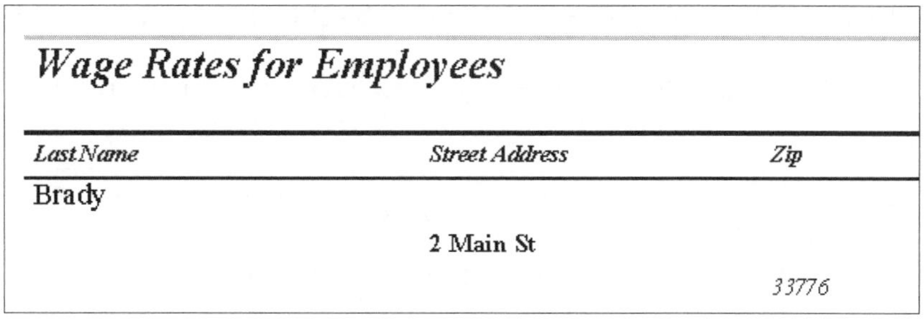

Wage Rates for Employees

LastName *Street Address* *Zip*

Brady

2 Main St

33776

Figure B-76 Wage Rate for Employees grouped report (Wage Rate not shown)

Suppose that you want the last name, street address, and zip all on the same line. The way to do that is to take the report into the Design View for editing. At the main objects menu, select "Wage Rate for Employees" Report and Design. At this point, the "headers" look like those shown in Figure B-77.

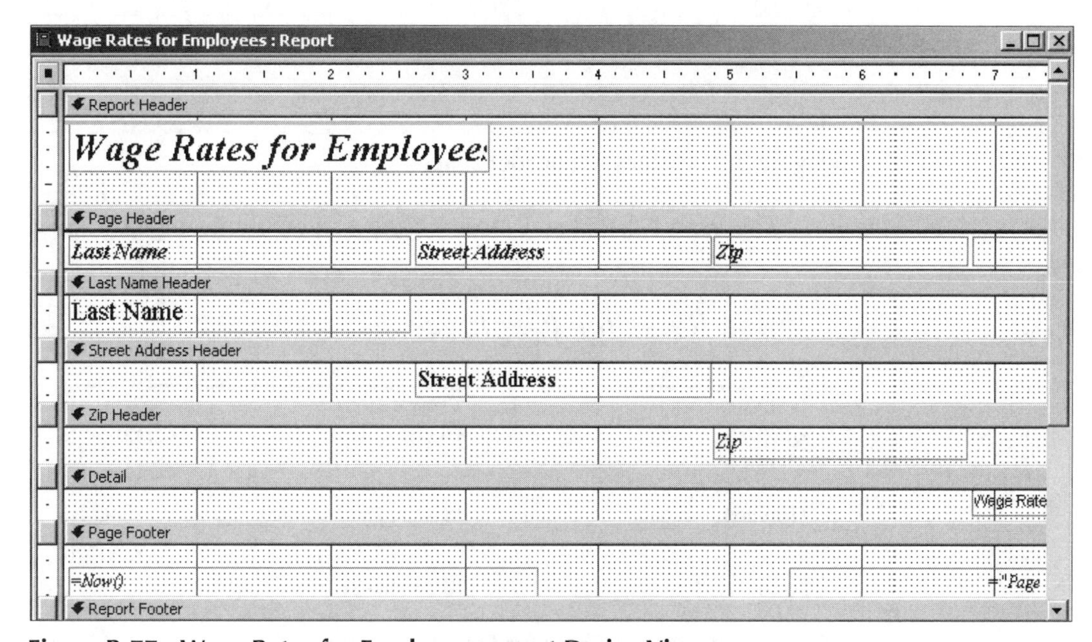

Figure B-77 Wage Rates for Employees report Design View

Your goal is to get the Street Address and Zip fields into the last name header (*not* into the page header!), so they will then print on the same line. The first step is to click the Street Address object in the Street Address Header as shown in Figure B-78.

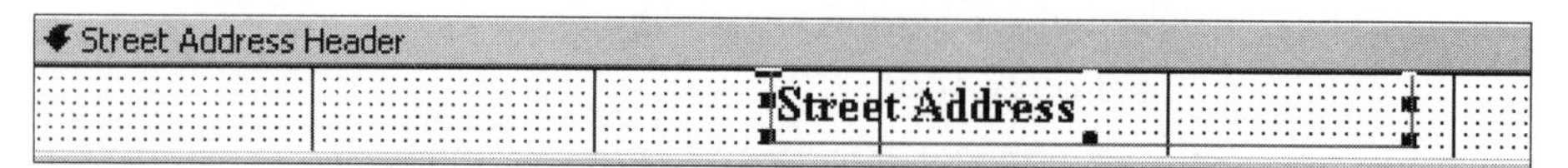

Figure B-78 Selecting Street Address object in the Street Address header

Hold the button down with the little "paw" icon and drag the object up into the Last Name Header as shown in Figure B-79.

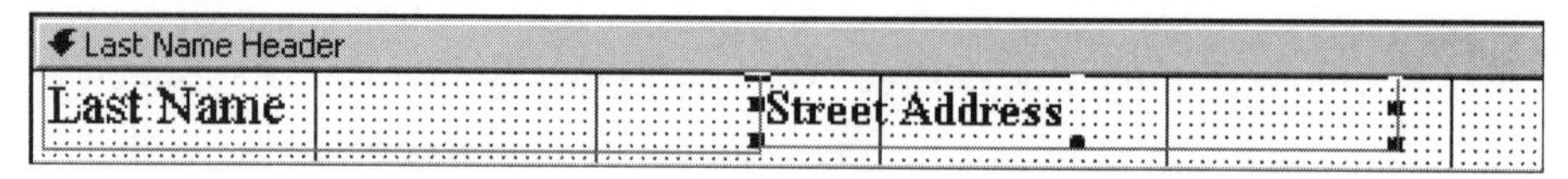

Figure B-79 Moving the Street Address object to the Last Name header

Do the same thing with the Zip object as shown in Figure B-80.

Figure B-80 Moving the Zip object to the Last Name header

To get rid of the header space allocated to the objects, tighten up the "dotted" area between each header. Put the cursor on the top of the header panel. The arrow changes to something that looks like a crossbar. Click and drag up to close the distance. After both headers are moved up, your screen should look like that shown in Figure B-81.

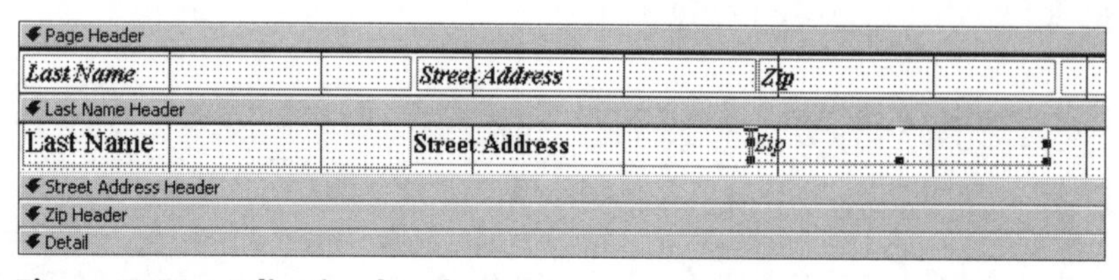

Figure B-81 Adjusting header space

Your report should now resemble the portion of the one shown in Figure B-82.

Wage Rates for Employees

Last Name	*Street Address*	*Zip*
Brady	2 Main St	33776
Howard	28 Sally Dr	19702
Jones	18 Spruce St	19716

Figure B-82 Wage Rates for Employees report

FORMS

Forms simplify adding new records to a table. The Form Wizard is easy to use and will not be reviewed; however, making a form out of more than one table needs to be explained.

When you base a form on one table, you simply identify that table when you are in the Form Wizard set-up. The form will have all the fields from that table and only those fields. When data is entered into the form, a complete new record is automatically added to the table.

But what if you need a form that includes the data from two (or more) tables? Begin (counterintuitively) with a query. Bring all tables you need in the form into the query. Bring down the fields you need from each table. (For data entry purposes, this probably means bringing down *all* the fields from each table.) All you are doing is selecting fields that you want to show up in the form, so you make *no criteria* after bringing fields down in the query. Save the query. When making the form, tell Access to base the form on the query. The form will show all the fields in the query; thus, you can enter data into all the tables at once.

Suppose that you want to make one form that would, at the same time, enter records into the EMPLOYEE table and the WAGE DATA table. The first table holds relatively permanent data about an employee. The second table holds data about the employee's starting wage rate, which will probably change.

The first step is to make a query based on both tables. Bring all the fields from both tables down into the lower area. Basically the query just gathers up all the fields from both tables into one place. No criteria is needed. Save the query.

The second step is to make a form based on the query. This works because the query knows about all the fields. Tell the form to display all fields in the query. (Common fields—here, SSN—would appear twice, once for each table.)

Tutorial B

IMPORTING DATA

Text or spreadsheet data is easily imported into Access. In business, importing data happens frequently due to disparate systems. Assume that your healthcare coverage data is on the Human Resources Manager's computer in an Excel spreadsheet. Open the software application Microsoft Excel. Create that spreadsheet in Excel now, using the data shown in Figure B-83.

	A	B	C
1	SSN	Provider	Level
2	114-11-2333	BlueCross	family
3	123-45-6789	BlueCross	family
4	148-90-1234	Coventry	spouse
5	222-82-1122	None	none
6	714-60-1927	Coventry	single
7	Your SSN	BlueCross	single

Figure B-83 Excel data

Save the file, then close it. Now you can easily import that spreadsheet data into a new table in Access. With your **Employee** database open and Tables object selected, click New and click Import Table, as shown in Figure B-84. Click OK.

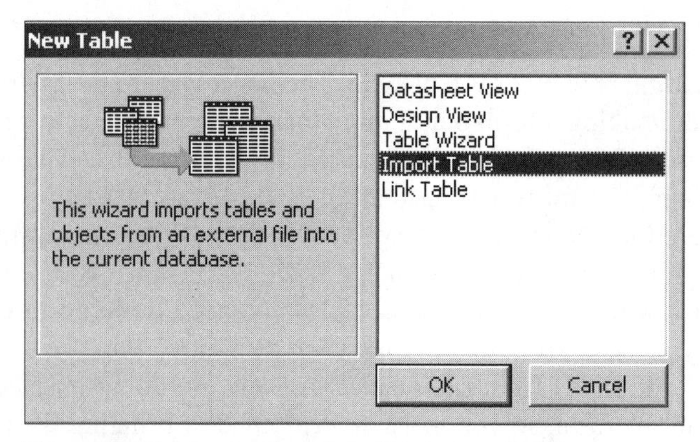

Figure B-84 Importing data into a new table

Find your spreadsheet. Be sure to choose **Microsoft Excel** as **Files of Type**. Assuming that you just have one worksheet in your Excel file, your next screen looks like that shown in Figure B-85.

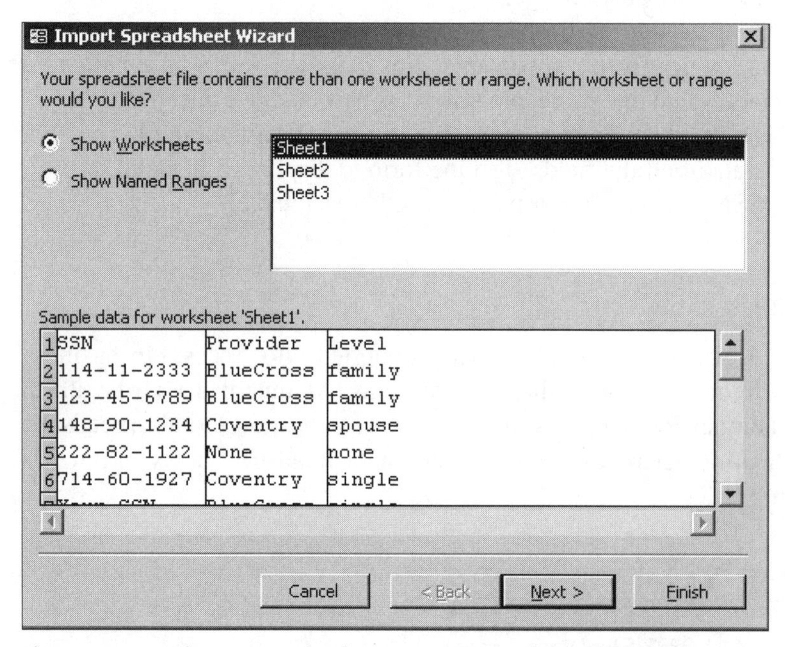

Figure B-85 First screen in the Import Spreadsheet Wizard

Choose Next, and then make sure you click the box that says First Row Contains Column Headings, as shown in Figure B-86.

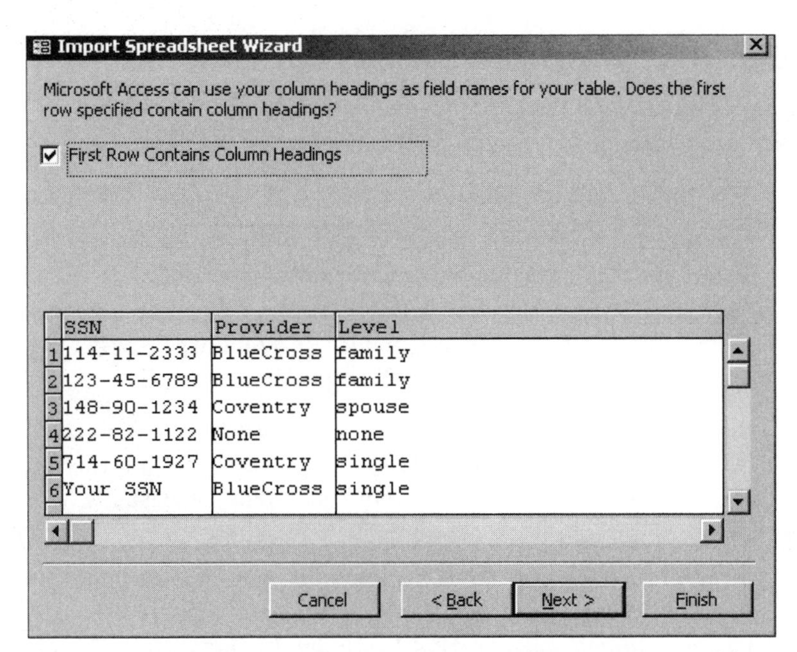

Figure B-86 Choosing column headings in the Import Spreadsheet Wizard

Store your data in a new table, do not index anything (next two screens of the Import Wizard), but choose your own primary key, which would be SSN, as chosen in Figure B-87.

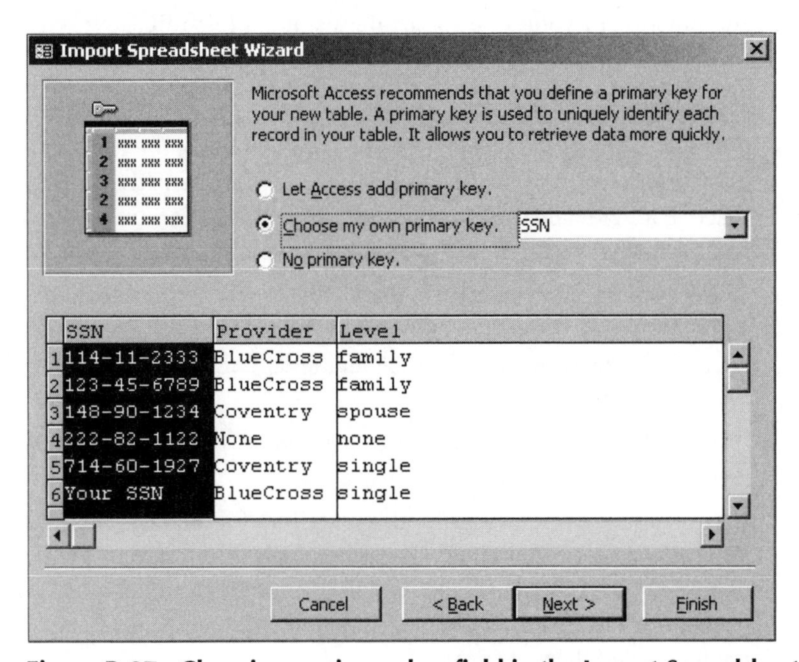

Figure B-87 Choosing a primary key field in the Import Spreadsheet Wizard

Continue through the Wizard, giving your table an appropriate name. After the table is imported, take a look at it and its design. (Highlight Table, Design button.) Note the width of each field (very large). Adjust the field properties as needed.

TROUBLE-SHOOTING COMMON PROBLEMS

Access beginners (and veterans!) sometimes create databases that have problems. Common problems are described here, along with their causes and corrections.

1. *"I saved my database file, but it is not on my diskette! Where is it?"*

   You saved to some fixed disk. Use the Find option of the Windows Start button. Search for all files ending in ".mdb" (search for *.mdb). If you did save it, it is on the hard **drive (C:\)** or on some network drive. (Your site assistant can tell you the drive designators.) Once you have found it, use Windows Explorer to copy it to your diskette in **drive A:**. Click it, and drag to **drive A:**.

   Reminder: Your first step with a new database should be to Open it on the intended drive, which is usually **drive A:** for a student. Don't rush this step. Get it right. Then, for each object made, save it *within* the current database file.

2. *"What is a 'duplicate key field value'? I'm trying to enter records into my Sales table. The first record was for a sale of product X to customer #101, and I was able to enter that one. But when I try to enter a second sale for customer #101, Access tells me I already have a record with that key field value. Am I only allowed to enter one sale per customer!?"*

   Your primary key field needs work. You may need a compound primary key—CUSTOMER NUMBER and some other field or fields. In this case, CUSTOMER NUMBER, PRODUCT NUMBER, and DATE OF SALE might provide a unique combination of values—or consider using an INVOICE NUMBER field as a key.

3. *"My query says 'Enter Parameter Value' when I run it. What is that?"*

   This symptom, 99 times out of 100, indicates you have an expression in a Criteria or a Calculated Field, and *you misspelled a field name in the expression.* Access is very fussy about spelling. For example, Access is case sensitive. Furthermore, if you put a space in a field name when you define the table, then you must put a space in the field name when you reference it in a query expression. Fix the typo in the query expression.

   This symptom infrequently appears when you have a calculated field in a query, and you elect *not* to show the value of the calculated field in the query output. (You clicked off the Show box for the calculated field.) To get around this problem, click Show back on.

4. *"I'm getting a fantastic number of rows in my query output—many times more than I need. Most of the rows are duplicates!"*

   This symptom is usually caused by a failure to link together all tables you brought into the top half of the query generator. The solution is to use the manual click-and-drag method. Link the fields (usually primary key fields) with common *values* between tables. (Spelling of the field names is irrelevant because the link fields need not be spelled the same.)

5. *"For the most part, my query output is what I expected, but I am getting one or two duplicate rows."*

   You may have linked too many fields between tables. Usually only a single link is needed between two tables. It's unnecessary to link each common field in all combinations of tables; usually it's enough to link the primary keys. A layman's explanation for why over-linking causes problems is that excess linking causes Access to "overthink" the problem and repeat itself in its answer.

On the other hand, you might be using too many tables in the query design. For example, you brought in a table, linked it on a common field with some other table, but then did not use the table. You brought down none of its fields and/or you used none of its fields in query expressions. Therefore, get rid of the table, and the query should still work. Try doing this to see whether the few duplicate rows disappear: Click the unneeded table's header in the top of the QBE area and press the Delete key.

6. *"I expected six rows in my query output, but I only got five. What happened to the other one?"*

   Usually this indicates a data-entry error in your tables. When you link together the proper tables and fields to make the query, remember that the linking operation joins records from the tables *on common values* (*equal* values in the two tables). For example, if a primary key in one table has the value "123", the primary key or the linking field in the other table should be the same to allow linking. Note that the text string "123" is not the same as the text string "123 " —the space in the second string is considered a character too! Access does not see unequal values as an error: Access moves on to consider the rest of the records in the table for linking. Solution: Look at the values entered into the linked fields in each table and fix any data entry errors.

7. *"I linked fields correctly in a query, but I'm getting the empty set in the output. All I get are the field name headings!"*

   You probably have zero common (equal) values in the linked fields. For example, suppose you are linking on Part Number (which you declared as text): in one field you have part numbers "001", "002", and "003", and in the other table part numbers "0001", "0002", and "0003". Your tables have no common values, which means no records are selected for output. You'll have to change the values in one of the tables.

8. *"I'm trying to count the number of today's sales orders. A Sigma query is called for. Sales are denoted by an invoice number, and I made this a text field in the table design. However, when I ask the Sigma query to 'Sum' the number of invoice numbers, Access tells me I cannot add them up! What is the problem?"*

   Text variables are words! You cannot add words, but you can count them. Use the Count Sigma operator (not the Sum operator): count the number of sales, each being denoted by an invoice number.

9. *"I'm doing Time arithmetic in a calculated field expression. I subtracted the Time In from the Time Out and I got a decimal number! I expected 8 hours, and I got the number .33333. Why?"*

   [Time Out] – [Time In] yields the decimal percentage of a 24-hour day. In your case, 8 hours is a third of a day. You must complete the expression by multiplying by 24: ([Time Out] – [Time In]) * 24. Don't forget the parentheses!

10. *"I formatted a calculated field for currency in the query generator, and the values did show as currency in the query output; however, the report based on the query output does not show the dollar sign in its output. What happened?"*

    Go into the report Design View. There is a box in one of the panels representing the calculated field's value. Click the box and drag to widen it. That should give Access enough room to show the dollar sign, as well as the number, in output.

11. *"I told the Report Wizard to fit all my output to one page. It does print to just one page. But some of the data is missing! What happened?"*

Access fits the output all on one page by *leaving data out*! If you can stand to see the output on more than one page, click off the "Fit to a Page" option in the Wizard. One way to tighten output is to go into the Design View and remove space from each of the boxes representing output values and labels. Access usually provides more space than needed.

12. *"I grouped on three fields in the Report Wizard, and the Wizard prints the output in a staircase fashion. I want the grouping fields to be on one line! How can I do that?"*

    Make adjustments in the Design View. See the Reports section of this tutorial for instruction.

13. *"When I create an Update query, Access tells me that zero rows are updating, or more rows are updating than I want. What is wrong?"*

    If your Update query is not correctly set up, for example, if the tables are not joined properly, it will either try not to update anything, or it will update all the records. Check the query, make corrections, and run it again.

14. *"After making a Summation Query with a Sum in the Group By row and saving that query, when I go back to it, the Sum field now says Expression, and Sum is put in the field name box. Is this wrong?"*

    Access sometimes changes that particular statistic when the query is saved. The data remains the same, and you can be assured your query is correct.

Preliminary Case: The Mountain-Biking Club

1
CASE

Setting Up a Relational Database to Create Tables, Queries, and Reports

Preview

In this case, you'll create a relational database for a mountain-biking club that holds events for members throughout the year. First, you'll create four tables and populate them with data. Next, you'll create two queries: a Parameter query and a Count query. Finally, you'll create a report that shows members' participation in yearly events.

Preparation

- Before attempting this case, you should have experience using Microsoft Access.
- Complete any part of the previous tutorials that your instructor assigns, or refer to them as necessary.

Background

A local mountain-biking club has organized a database to keep track of members, what type of bike(s) each rides, available events, and member participation. The secretary of the club created the database tables in Access and populated them with data.

There are four tables in the database. One table, the MEMBERS table, contains the fields for Member ID, Last Name, First Name, Address, Telephone, and Date Joined—the date on which a member joined the club. A second table, the BIKES table, contains information about members' bikes. Note that some members have more than one bike, hence a one-to-many relationship, so the information about members' bikes is put in a separate table, rather than in the MEMBERS table. A third table, the EVENTS table, contains information about the club's events. A fourth table, the PARTICIPATION table, which is a transaction table, registers each member for the events in which he or she will ride.

The president of the bike club would like you to help him to use the database to perform several tasks. First, he wants to be able to see which members ride which type of bike. This information comes in handy when soliciting sponsorship for events. He would like to run the query and input just the bike type, and then see output as members' names and telephone numbers.

In addition, there are times when the president wants to be able to get an update of the number of club members. This information is passed on to the treasurer, who reconciles that count with dues paid, so this query should count the number of members.

Finally, the club officers need a report listing each event for the year and which members attended or are scheduled to attend.

Assignment 1 Creating Tables

Use Microsoft Access to create the tables with the fields shown in Figures 1-1 through 1-4 and discussed in the Background section. Populate the database tables as shown. Add your name to the MEMBERS table as Member ID 10006, complete with your address and telephone number. (*Note*: Your Date Joined will remain as shown in Figure 1-1.)

Members : Table

Member ID	Last Name	First Name	Address	Telephone	Date Joined
10001	Monk	Ellen	009 Purnell	831-1794	2/1/1998
10002	Brady	Joseph	008 Purnell	831-1765	3/4/1997
10003	Maida	Claude	234 Main	737-0000	12/31/2001
10004	Smith	Sally	34 East 1st St	737-1111	6/7/1998
10005	Jones	Sammy	14 Fifth St	737-9999	9/15/1998
10006	Yours	Yours	Yours	Yours	1/1/2003

Figure 1-1 The MEMBERS table

Bikes : Table

Bike ID	Member ID	Bike Type
101	10001	Orange P7
102	10001	Diamondback X-Link
103	10002	Kona Lava Dome
104	10003	Kona Nunu
105	10004	Kona Lava Dome
106	10004	Diamondback X-Link
107	10005	Kona Lava Dome
108	10006	Orange P7

Figure 1-2 The BIKES table

Events : Table

Event ID	Level	Location	Date	Start Time	End Time
00010	Easy	Fair Hill	9/3/2004	10:00:00 AM	2:00:00 PM
00011	Intermediate	White Clay	9/10/2004	3:00:00 PM	5:00:00 PM
00012	Advanced	Poconos	11/4/2004	9:30:00 AM	9:00:00 PM
00013	Intermediate	Spruce Run	1/20/2005	11:00:00 AM	4:00:00 PM
00014	Intermediate	Rickets Glen	2/14/2005	9:30:00 AM	3:30:00 PM
00015	Intermediate	Willow Falls	3/18/2005	12:00:00 PM	5:00:00 PM
00016	Easy	Fair Hill	4/1/2005	10:00:00 AM	11:30:00 AM
00017	Easy	Fair Hill	5/4/2005	10:00:00 AM	11:30:00 AM
00018	Advanced	Middle Run	6/6/2005	9:00:00 AM	5:00:00 PM
00019	Easy	Fair Hill	7/4/2005	10:00:00 AM	11:30:00 AM
00020	Intermediate	Rickets Glen	8/16/2005	9:00:00 AM	5:00:00 PM
00021	Advanced	Spruce Run	9/2/2005	8:30:00 AM	4:00:00 PM

Figure 1-3 The EVENTS table

Participation : Table

Event ID	Member ID
00010	10001
00010	10003
00010	10005
00010	10006
00011	10002
00011	10004
00011	10005
00011	10006
00012	10001
00012	10004
00012	10005
00012	10006
00013	10001
00013	10006
00014	10002
00014	10006

Figure 1-4 The PARTICIPATION table

ASSIGNMENT 2 CREATING QUERIES AND REPORTS

Assignment 2A: Creating a Parameter Query

Create a Parameter query that prompts the user for the names of the members who own a specific type of bike. Include only the fields Last Name, First Name, and Telephone in the output. Example: If the club president used the query to find the owners of Orange P7 bikes, the output would resemble that shown in Figure 1-5.

Owners of Bike Type? : Select Query

Last Name	First Name	Telephone
Monk	Ellen	831-1794
Yours	Yours	Yours

Figure 1-5 Query: Owners of Bike Type?

Run the Parameter query with the input of Orange P7. Save the query as Owners of Bike Type? and print the results.

Assignment 2B: Creating a Count Query

Create a query that counts the number of club members. Your output should resemble that shown in Figure 1-6. Print the output.

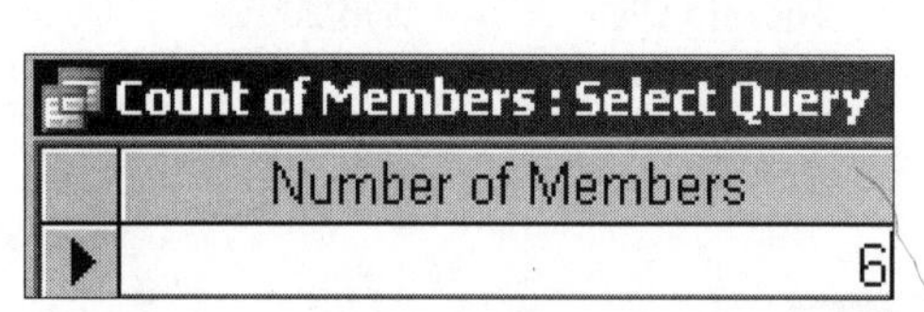

Count of Members : Select Query

Number of Members
6

Figure 1-6 Query: Count of Members

Note the heading change (Number of Members). Save the query as Count of Members. Print the output.

Assignment 2C: Generating a Yearly Participation Report

Generate a report that shows the participation of each member in every bike event during the year. To create the report, you will need to do the following:

- You must first create the query that will bring all the data together for the report.
- In the Report Wizard, bring in the following fields: Location, Date, Last Name, First Name, and Telephone.
- Group on Date and Location. Sort on Date.
- You will find that the Date field is in the Detail Band of the report design. That will have to be moved to the Location Header Band.
- Use Print Preview to make sure the report will print correctly.
- You will have to adjust the report design in Design View so the report resembles the portion of the report shown in Figure 1-7 (with other adjustments to make the report look correct).
- Save the report as Yearly Participation.

Yearly Participation					
Date by Month	***Location***	***Date***	***Last Name***	***FirstName***	***Telephone***
September 2004					
	Fair Hill	9/3/2004			
			Yours	Yours	Yours
			Jones	Sammy	737-9999
			Maida	Claude	737-0000
			Monk	Ellen	831-1794
	White Clay	9/10/2004			
			Yours	Yours	Yours
			Jones	Sammy	737-9999
			Smith	Sally	737-1111
			Brady	Joseph	831-1765
November 2004					
	Poconos	11/4/2004			
			Yours	Yours	Yours
			Jones	Sammy	737-9999
			Smith	Sally	737-1111

Figure 1-7 Report: Yearly Participation (portion of report)

Make sure you close the database file before removing your diskette.

DELIVERABLES

Assemble the following deliverables for your instructor:

1. Printouts of four tables
2. Query output: Owners of Bike Type?
3. Query output: Count of Members
4. Report: Yearly Participation
5. Diskette with database file

Staple all pages together. Put your name and class number at the top of the page. Make sure your diskette is labeled.

2 CASE

The Equestrian Club

Designing a Relational Database to Create Tables, Queries, and Reports

Preview

In this case, you'll design a relational database for the Equestrian Club at your university. After your database design is completed and correct, you will create database tables and populate them with data. Then you will produce three queries and one report. The queries will answer these questions: Which members have had their physical exam? Which members are riding the most? In which events is a specific member involved? The report will list member participation in each event.

Preparation

- Before attempting this case, you should have some experience in database design and in using Microsoft Access.
- Complete any part of Database Design Tutorial A that your instructor assigns.
- Complete any part of Access Tutorial B that your instructor assigns, or refer to the tutorial as necessary.
- Refer to Tutorial E as necessary.

BACKGROUND

You have been elected secretary of the Equestrian Club at your university. The previous secretary kept all records by hand, but because you have experience in database design and Microsoft Access, you feel that it is time to computerize the club's information.

You have several goals that you want to accomplish with the database. First, you need to keep track of all the current club members—their member ID number, address, and telephone number. It also is important to note whether a member has had a yearly physical examination. (University regulations require that all students who participate in a sports-related club have a physical examination at least once a year.)

In addition, the club sponsors a number of events throughout the academic year. Each event has three competitions: dressage, stadium jumping, and cross-country riding. Those events are held locally and are scheduled for a specific date with specific start and end times, and each event has a unique event ID number. Thus, another of your tasks is to keep track of which members are signed up to attend each event.

Also, to comply with university regulations regarding physical examinations, you must keep track of which members have not had their annual physical and send out reminders to those members.

Some members ride more than others, and the officers like to reward the club's top riders. Thus, another piece of information that should be gleaned from a query is a list of riders, organized by those who have the most riding time to those with the least riding time. As secretary, you also need to have up-to-the-minute information about each event in which each participant is riding. It would be nice to be able to type in a member's name and see the events in which that member participated.

Your last task is to prepare a report. Each month the secretary is required to generate a report for the other club officers that lists each event and the names of the participants in each event.

ASSIGNMENT 1 CREATING THE DATABASE DESIGN

In this assignment, you will design your database tables on paper, using a word-processing program. Pay close attention to the tables' logic and structure. Do not start your Access code (Assignment 2) before getting feedback from your instructor on Assignment 1. Keep in mind that you will need to look at what is required in Assignment 2 to design your fields and tables properly. It's good programming practice to look at the required outputs before designing your database. When designing the database, observe the following guidelines:

- First, determine the tables you'll need by listing on paper the name of each table and the fields that it should contain. Avoid data redundancy. Do not create a field if it could be created by a "calculated field" in a query.
- Include a logical field that answers the question "Yearly Physical?"
- You'll need a transaction table. Avoid duplicating data.
- Document your tables by using the Table facility of your word processor. Your word-processed tables should resemble the format of the table in Figure 2-1.
- You must mark the appropriate key field(s). You can designate a key field by an asterisk (*) next to the field name. Keep in mind that some tables need a compound primary key to identify a record uniquely within a table.
- Print out the database design.

Table Name	
Field Name	**Data Type (text, numeric, currency, etc.)**
...	...
...	...

Figure 2-1 Table design

Have your design approved before beginning Assignment 2; otherwise, you may need to redo Assignment 2.

ASSIGNMENT 2 CREATING THE DATABASE AND MAKING QUERIES AND A REPORT

In this assignment, you will first create database tables in Access and populate them with data. Next, you will create three queries and a report.

Assignment 2A: Creating Tables in Access

In this part of the assignment, you will create your tables in Access. Use the following guidelines:

- Type records into the tables, using the students' names and addresses shown in Figure 2-2. Add your name and address as an additional student. You have not had your annual physical.
- Invent member IDs and telephone numbers. Select one student NOT to have had a yearly physical (in addition to yourself).
- Assume that there are four events on different dates, with differing start and end times. Name the events and the start and end time for each; choose any location for the events. Give each event a unique ID number.
- Have each club member participate in at least one event. Make one member participate in at least two events, and have one member participate in three events.
- Appropriately limit the size of the text fields; for example, a telephone number does not need to be the default setting of 50 characters in length.
- Print all tables.

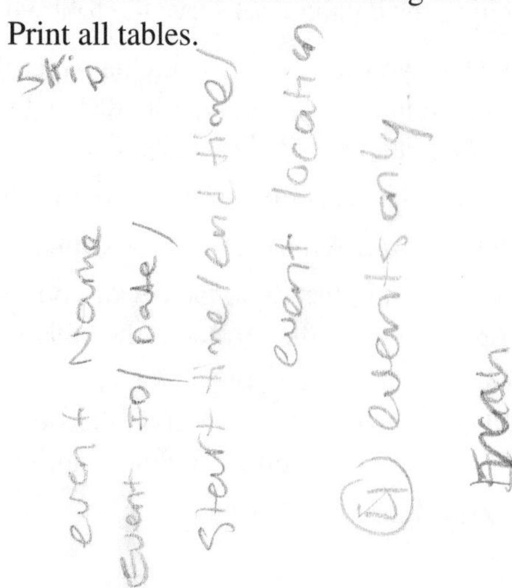

Last Name	First Name	Local Address
Franklin	John	345 Russell Hall
McGregor	Taylor	123 Lane Hall
Thompson	Michael	219 Smyth Hall
Mason	Lee	423 Rodney Hall
Falkner	Nicole	333 Rodney Hall
O'Shea	Jamie	109 Lane Hall
Your last name	Your first name	Your address

Figure 2-2 Data

Assignment 2B: Creating Queries and a Report

There are three queries and one report to generate, as outlined in the Background section.

Query 1

Create a query called No Physicals. The output of the query should list only the last name, first name, address, and telephone number of only those students who have NOT had their yearly physicals. Your output should resemble that shown in Figure 2-3.

No Physicals : Select Query

Member Last Name	Member First Name	Address	Phone
Falkner	Nicole	333 Rodney Hall	302-837-0123
Your name	Your name	Your address	Your tel

Figure 2-3 Query: No Physicals (example output)

Query 2

Create a query called Total Time. When creating the query, note the following:

- Show each member's last name and first name and total time participating in all events.
- List the members who ride in order, ranking them from those who ride the most to those who ride the least.
- Your data will differ, but your output should resemble the format shown in Figure 2-4.

Total Time : Select Query

Member Last Name	Member First Name	Total Time
Your name	Your name	10.25
Franklin	John	6.5
O'Shea	Jamie	5
Falkner	Nicole	5
Mason	Lee	4
McGregor	Taylor	2.5

Figure 2-4 Query: Total Time (example output)

Query 3

Create a query called Member's Events. When creating the query, note the following:

- There should be a Parameter query that will give information on each student's events.
- The query should prompt the user for Member Last Name input.
- The output should show the member's last name, event ID, date, and total time for the participant for that event.
- Your data may differ, but if you use the last name "O'Shea," your output should resemble the format shown in Figure 2-5.

Member's Events : Select Query

Member Last Name	Event ID	Date	Event Time
O'Shea	10102	10/1/2004	5

Figure 2-5 Query: Member's Events (example output)

Report

Create a report called Participant List. Your report's output should show headers for Event ID, Event Location, Member Last Name, and Member First Name. Use the following procedure:

1. First, create a query for input to that report.
2. Group on the fields Event ID and Event Location.
3. Adjust the design of the report to align data under the Event ID and Event Location headers.
4. Depending on your data, your output should resemble that shown in Figure 2-6.

Participant List

Event ID	*Event Location*	*Member LastName*	*Member First Name*
10101	*Radnor, PA*		
		Your name	Your name
		McGregor	Taylor
		Franklin	John
10102	*Elkton, MD*		
		O'Shea	Jamie
		Falkner	Nicole
10103	*Lancaster, PA*		
		Your name	Your name
10104	*Newark, DE*		
		Your name	Your name
		Mason	Lee
		Franklin	John

Figure 2-6 Report: Participant List (example output)

ASSIGNMENT 3 MAKING A PRESENTATION

Create a presentation that explains the database to your fellow club officers. Include the design of your database tables and how to use the database. Your presentation should take fewer than 10 minutes, including a brief question-and-answer period.

DELIVERABLES

Case 2

Assemble the following deliverables for your instructor:

1. Word-processed design of tables
2. Tables created in Access
3. Query 1: No Physicals
4. Query 2: Total Time
5. Query 3: Member's Events (with one last name)
6. Report: Participant List
7. Presentation materials
8. Any other required tutorial printouts or tutorial diskette

Staple all pages together. Put your name and class number at the top of the page. Make sure your diskette is labeled.

3

CASE

The Yoga Class

Designing a Relational Database to Create Tables, Queries, Forms, and Reports

Preview

In this case, you'll design a relational database for a yoga specialist who offers classes in his home studio. After your database design is completed and correct, you will create database tables and populate them with data. Next, you will create a form for data input. Then you'll create four queries: one to output the number of weeks in each class, one to count the number of students in each class, a Delete query to delete those students who have not paid their tuition, and a Parameter query to generate a class list. Finally, you'll create a report that displays tuition fees for all students.

Preparation

- Before attempting this case, you should have experience in database design and in using Microsoft Access.
- Complete any part of Tutorial A that your instructor assigns.
- Complete any part of Tutorial B that your instructor assigns, or refer to the tutorial as necessary.
- Refer to Tutorial E as necessary.

➤ BACKGROUND

You've begun taking classes with a local yoga specialist who teaches yoga in his home studio. The specialist has been teaching yoga for years, and with the recent surge in yoga's popularity his business has grown so much that he has hired instructors to help him teach the many classes he offers. The logistics of teaching classes and keeping records by hand has become too cumbersome. So, knowing that you are an Access expert, the business owner has asked you to create a database for him. In addition to having you create the database, he wants you to create a form, some queries, and a report that he can use for record keeping.

The yoga classes have quite a loyal following, and most students sign up for classes again and again. So, the business owner wants to keep student data—for past and current students—in the database. At present, he writes student data on an index card when a student registers to take a class. On each card, he notes the student's name, address, date of birth, and gender. After some mix-ups with three students having the same last name, he's decided that each student should also have a unique student ID number. He then notes on each card whether the student has paid for the class. All this information must be captured in the database, and it can be input via a form to make it easy for the business owner.

The yoga classes are quite varied. Some yoga classes cost more than others, depending on the focus of the class and the teacher's expertise. Some classes have more weekly sessions than others. The times at which classes begin and end are staggered throughout the day and evening. All these details need to be tracked in the database. To simplify things, the business owner wants each class to have a unique ID number in addition to its name. He also needs to keep track of his instructors and the class or classes that each instructor teaches.

The business owner would like you to create a number of queries. He frequently receives telephone calls from people who want to know the duration (in weeks) of a particular class. He'd like to run a query so he can quickly answer that question. He also gets telephone calls from people who want to know whether they can get into a particular class. He has no way of knowing whether a class is filled, so he needs to be able to run a query to find out how many students are in each class. The business owner also wants to increase his class offerings, so he'd like to be able to run a query that will tell him which classes are the most popular.

Of course, this is a money-making business, and the business owner needs to collect tuition. If a student does not pay for a class by the third week, the business owner tells the delinquent student that he or she is no longer enrolled. The database must mirror the business owner's actions, so he would like a query that will delete those students who have not paid their tuition and have been "dis-enrolled."

The business owner also needs to provide his instructors with a list of students in their classes. You can create a Parameter query that will allow the business owner to type in the name of the class and see a list of the students' names.

Because the business owner also keeps the books for his business, he would like a report that lists each student's tuition charges.

➤ ASSIGNMENT 1 CREATING THE DATABASE DESIGN

In this assignment, you will design your database tables on paper, using a word-processing program. Pay close attention to the tables' logic and structure. Do not start your Access code (Assignment 2) before getting feedback from your instructor on Assignment 1. Keep in mind that you will need to look at what is required in Assignment 2 to design your fields and tables properly. It's good programming practice to look at the required outputs before designing your database. When designing the database, observe the following guidelines:

- First, determine the tables you'll need by listing on paper the name of each table and the fields that it should contain. Avoid data redundancy. Do not create a field if it could be created by a "calculated field" in a query.
- You'll need a transaction table. Avoid duplicating data.
- Include a logical field that answers the question, "Tuition paid?"
- Document your tables by using the Table facility of your word processor. Your word-processed tables should resemble the format of the table in Figure 3-1.
- You must mark the appropriate key field(s). You can designate a key field by an asterisk (*) next to the field name. Keep in mind that some tables need a compound primary key.
- Print out the database design.

Table Name	
Field Name	**Data Type (text, numeric, currency, etc.)**
...	...
...	...

Figure 3-1 Table design

Have your design approved before beginning Assignment 2; otherwise, you may need to redo Assignment 2.

Assignment 2 Creating a Database and Developing a Form, Queries, and a Report

In this assignment, you will first create database tables in Access and populate them with data. Next, you will create one form, four queries, and a report.

Assignment 2A: Creating Tables in Access

In this part of the assignment, you will create your tables in Access. Observe the following guidelines:

- Type records into the tables, using the students shown in Figure 3-2. Add your name and address as an additional student. Use your own hometown data for the City and State fields.
- Make up dates of birth for each student.
- Assume that there are at least six classes. You can use some of the following class names for types of yoga: Ananda, Bikram, Kripalu, and Iyengar. Each class has a specific tuition, starts on a particular date, ends on another date, is at a specific time, and is on a specific day of the week. Make up all this data.

- There should be at least three instructors teaching the classes. Make up this data.
- Sign up students for classes. Have each student attend at least one class and have three of them attend two classes. Designate one student as "not paid."
- Appropriately limit the size of the text fields; for example, a class ID does not need to be the default setting of 50 characters in length.
- Print all tables.

Last Name	First Name	Permanent Address
Fischer	Patti	1900 South Chapel Street
DeVries	Marcus	101 Pinecone Street
Puig	Martha	99 Beech Street
Greaney	Petra	98 College Street
Brown	Robert	29 Hickory Lane
Sanchez	Maria	9 Spruce Street
Your last name	Your first name	Your address

Figure 3-2 Student data

Assignment 2B: Creating a Form, Queries, and a Report

You'll need to create one form, four queries, and one report, as outlined in the Background section of this case.

Form

Create a form that the business owner can use to type in the class registration. Base this form on the registration information only, and use the Form Wizard. Save the form as Data Entry. Print one record from this form.

Query 1

The yoga specialist would like to advertise the length of each class (in weeks). Create a query to gather this information. When creating the query, note the following:

- Your query should show the class ID, the class name, and the number of weeks that the class meets.
- Call the query Class Weeks.

Your data will differ, but your output should resemble the format shown in Figure 3-3.

Class Weeks : Select Query

Class ID	Class Name	Length of Class (weeks)
101	Ananda	7
102	Bikram	2
103	Kripalu	7
104	Iyengar	7

Figure 3-3 Query: Class Weeks (example output)

Query 2

For accounting purposes, the yoga specialist would like to know how many students are enrolled in each class, from the most populated class to the least populated. Create a query to produce this information. When creating the query, note the following:

- Your query should show the class name and the number of students.
- Change any column headings, if necessary.
- Call the query Number of Students in Classes.
- Think about what type of query you need to make.
- Sort this query's output.

Your data will differ, but your output should resemble the format shown in Figure 3-4.

Number of Students in Classes : Select Query

Class Name	Number of Students
Iyengar	4
Bikram	4
Ananda	2
Kripalu	1

Figure 3-4 Query: Number of Students in Classes (example output)

Query 3

Create a Delete query to remove from the database the students who have not paid their tuition. Any such record needs to be deleted from the table that contains registration data. Note the following:

- When you run the query, you should see a dialog box that warns of one record's being deleted (assuming that only one student did not pay).
- Delete that record.
- Save the query as Delete Pupils Not Paying.
- Print the table that contains registration data after deleting the record. Write on the top: "Updated Registration Table."

Query 4

Make a Parameter query to print a class list of students. When you create the query, observe the following requirements:

- Show class ID number, class name, student last name, and student first name.
- Allow the user to type in the name of the class and see the subsequent class list, i.e., prompt for the name of a class.
- Save the query as Class List—Parameter.

If you typed in "Ananda" for the class name, and depending on the records in your database, you should see output similar to that shown in Figure 3-5.

Class List - Parameter : Select Query			
Class ID	Class Name	Last Name	First Name
101	Ananda	Greaney	Petra
101	Ananda	Brown	Robert

Figure 3-5 Query: Class List—Parameter (example output)

Report

Create a report named Students' Tuition, using the following guidelines:

- First, create a query that displays how much each student has paid for classes thus far.
- Show the student's last name, first name, class ID number, class name, and price for each class.
- Bring the query into the Report Wizard and group on the students' names.
- Click the Summary Options button to sum the total fee for each student.
- Move the First Name field data into the student's Last Name header so the report looks good.
- Delete any extraneous total lines and sums.

Your data will differ, but the top of your output should resemble the format shown in Figure 3-6.

Students' Tuition

LastName	*First Name*	*Class ID*	*Class Name*	*Price*
Brown	Robert			
		104	Iyengar	$55.00
		102	Bikram	$90.00
		101	Ananda	$150.00
Total Owed				*$295.00*
DeVries	Marcus			
		104	Iyengar	$55.00
		102	Bikram	$90.00
Total Owed				*$145.00*

Figure 3-6 Report: Students' Tuition (example output)

ASSIGNMENT 3 GIVING A PRESENTATION

Create a presentation that explains the database to the yoga specialist and makes suggestions on ways to expand the business. Include the following:

- Describe the design of your database tables.
- Tell how to use the database, enter information into the form, and run the queries and reports.

- Do some research on the Internet and create a proposal for how to expand the business.
- Use the student data and Excel to create a profile of those students who might have the extra tuition money to pay for one-on-one instruction. Make a recommendation to the yoga specialist.
- If possible, include some graphics in your presentation to make it more interesting.

DELIVERABLES

Assemble the following deliverables for your instructor:

1. Word-processed design of tables
2. Tables created in Access
3. Form: Data Entry (one record printed only)
4. Query 1: Class Weeks
5. Query 2: Number of Students in Classes
6. Query 3: Printout of the Updated Registration Table
7. Query 4: Class List—Parameter
8. Report: Students' Tuition
9. Presentation materials
10. Any other required tutorial printouts or tutorial diskette

Staple all pages together. Put your name and class number at the top of the first page. Make sure your diskette is labeled.

The Animal Rehabilitation Clinic

4
CASE

Designing a Relational Database to Create Tables, Forms, Queries, and Reports

Preview

In this case, you'll design a relational database for an animal rehabilitation clinic. After your database design is completed and correct, you will create database tables and populate them with data. Then you'll create a form for data input to log animals arriving at the clinic. Next, you will create four queries: one to list the animal arriving on a specific date; two for calculating the length of stays (one by animal description and one by status description); and one for counting the number of different animals. Finally, you will create a summary report for the local animal control authorities.

Preparation

- Before attempting this case, you should have experience in database design and in using Microsoft Access.
- Complete any part of Tutorial A that your instructor assigns.
- Complete any part of Tutorial B that your instructor assigns, or refer to the tutorial as necessary.
- Refer to Tutorial E as necessary.

BACKGROUND

Mary Sweeney has an animal rehabilitation clinic in the basement of her home. She became interested in animal rehabilitation two years ago, took the proper training, and now takes in small animals that have been injured or abandoned and then nurses them back to health. Thus far, Mary has treated only wild birds, domestic birds, squirrels, raccoons, and ground hogs.

City animal control authorities are now asking Mary to complete many reports about her animals. She is overwhelmed by the workload. Not only does she often stay up all night feeding baby birds and such, but now she is required to keep meticulous records for the city. She asks for your help in setting up a database to keep records more easily.

Mary's record-keeping needs are not complicated. She needs to record each animal as it comes into her clinic. She gives each animal a unique number, records its animal type, and notes its status (i.e., oil spill, injury (misc.), car accident, or orphaned). She also records the date on which she releases the animal into the wild or finds a new home for it. Occasionally, some animals arrive dead (e.g., a person picks up an injured baby bird and puts it in a bucket, but it dies on the drive to Mary's house). Mary needs to record that such animals have entered the clinic but were "dead on arrival" (DOA).

City animal control officials require Mary to provide on a monthly basis the following four reports:

1. A list of all animals arriving on a specific date (city specifies the date)
2. A list of the animals' length of hospitalization, grouped by type of animal
3. A list of the animals' length of hospitalization, grouped by injury status
4. A count of animals, grouped by type of animal

You will need to create four queries that generate the information for each of those reports. In addition, Mary also needs a report that shows how many days each type of animal has spent at the clinic. An example of the typewritten report that she now uses is shown in Figure 4-1.

Final Report

Description	*Animal ID*	*Status Description*	*Total Days*
Ground Hog			
	101	Injured - Misc	62
Total Days			*62*
Raccoon			
	109	Injured - Misc	29
Total Days			*29*
Squirrel			
	103	Orphan	118
	102	Car Accident	93
Total Days			*211*

Figure 4-1 Example of report data

In addition, Mary would like you to create a form that would be filled out by people who drop off injured animals. The form should ask for the type of animal, the animal's status, and the date the animal arrived.

Assignment 1 Creating the Database Design

In this assignment, you will design your database tables on paper, using a word-processing program. Pay close attention to the tables' logic and structure. Do not start your Access code (Assignment 2) before getting feedback from your instructor on Assignment 1. To design your fields and tables properly, look at the output required in Assignment 2. It's good programming practice to look at the required outputs before designing your database. When designing the database, observe the following guidelines:

- First, determine the tables you'll need by listing on paper the name of each table and the fields that it should contain. Avoid data redundancy. Do not create a field if it could be created by a "calculated field" in a query.
- You'll need a transaction table. Avoid duplicating data.
- Include a logical field that answers the question, "DOA?"
- Create your tables by using the Table facility of your word processor. Your word-processed tables should resemble the format of the table in Figure 4-2.
- You must mark the appropriate key field(s). You can designate a key field by an asterisk (*) next to the field name. Keep in mind that some tables need a compound primary key.
- Print out the database design.

3 tables.

Case 4

Table Name	
Field Name	**Data Type (text, numeric, currency, etc.)**
...	...
...	...

Figure 4-2 Table design

NOTE

Have your design approved before beginning Assignment 2; otherwise, you may need to redo Assignment 2.

Assignment 2 Creating the Database and Developing a Form, Queries, and a Report

Assignment 2A: Creating Tables in Access

In this part of the assignment, you will create your tables in Access. Observe the following guidelines:

- Type records into the tables, using one of the following status codes: Car Accident (CA), Injured—Misc (IN), Orphan (OR), Oil Spill (OS).
- Thus far, the only animals Mary has taken in are domestic birds, ground hogs, raccoons, squirrels, and wild birds.

- Designate at least 10 animals (your choice of the ones mentioned in the previous step) to be at the clinic, some of which should arrive on the same day.
- Appropriately limit the size of the text fields; for example, an animal ID does not need to be the default setting of 50 characters in length.
- Print all tables.

Assignment 2B: Creating a Form, Queries, and a Report

You'll need to create one form, four queries, and one report, as outlined in the Background section of this case.

Form

Mary wants to have a form so that when people drop off injured animals, she can put their data directly into the Access database. When you create the form, do the following:

- Base the form on your transaction table.
- Make sure that when new data is typed into this form, the "date released" will be left blank.
- Print one record from the form.

Query 1

For this query, you need to list all the animals entering the clinic on a given date. (The user will be prompted for the date requested.) Note the following:

- The output of the query should show Animal ID, Description, Status Description, and Date In.
- You will need to create a Parameter query.
- If you were to run the query and enter the date 3/4/2005, your format would resemble that shown in Figure 4-3.
- Save the query as Entering Animals.

Entering Animals : Select Query

Animal ID	Description	Status Description	Date In
101	Ground Hog	Injured - Misc	3/4/2005
102	Squirrel	Car Accident	3/4/2005

Figure 4-3 Query: Entering Animals (example output)

Query 2

For this query, you need to calculate the number of days that all the animals in each animal group have stayed at the clinic. Use the following guidelines:

- List the description of the kind of animal and then rank the animals in descending order according to the number of days that all the animals in each group have stayed at the clinic.
- Your output headings should be Description and Total Length of Stay in Days.
- Make sure you have changed the column headings, if necessary.
- Save the query as Length of Stay vs. Type. (*Note*: You will not be able to add a period after "vs." because Access does not allow periods in a query name.)
- Your data will differ, but the format of your output should resemble that shown in Figure 4-4.

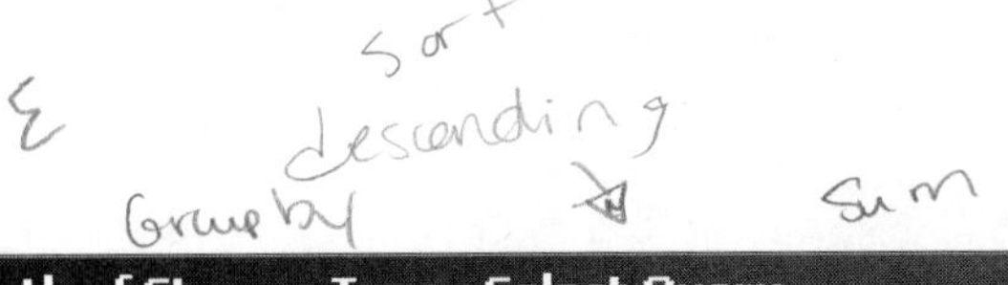

Length of Stay vs Type : Select Query	
Description	**Total Length of Stay in Days**
Wild Bird	379
Squirrel	211
Ground Hog	62
Raccoon	29

Figure 4-4 Query: Length of Stay vs. Type (example output)

Query 3

This query is similar to Query 2 except that you need to show the length of stay versus the status description. Once again, sort using a descending order based on the number of days stayed. Save the query as Length of Stay vs. Status, as shown in Figure 4-5.

Length of Stay vs Status : Select Query	
Status Description	**Total Length of Stay in Days**
Orphan	449
Car Accident	93
Injured - Misc	91
Oil Spill	48

Figure 4-5 Query: Length of Stay vs. Status (example output)

Query 4

For this query, you need to count the number of animals in each animal type. Observe the following requirements:

- Your output should show Description and Total Number of Animals.
- Sort in a descending order on Total Number of Animals.
- Make sure you create the correct type of query.
- Save your query as Number of Animals.
- Your data will differ, but your output should resemble that shown in Figure 4-6.

Number of Animals : Select Query	
Description	**Total Number of Animals**
Wild Bird	5
Squirrel	2
Raccoon	1
Ground Hog	1

Figure 4-6 Query: Number of Animals (example output)

Case 4

Report

Finally, you need to create a summary report. Your report should reflect the following requirements:

- It should show the kind of animal—the Description, Animal ID, Status Description, and Total Days that each animal has spent in the clinic.
- You will need to make a query to calculate the total number of days that the animals have spent in the clinic.
- Use the Report Wizard and group on the animal Description.
- Click the Summary Options button and total the days stayed for each animal description.
- Delete any unusual totals or lines. In Design View, correct any headings or titles as necessary.
- Your data will vary, but the format of your report should resemble that shown in the portion of the report in Figure 4-7.

Final Report

Description	*Animal ID*	*Status Description*	*Total Days*
Ground Hog			
	101	Injured - Misc	62
Total Days			*62*
Raccoon			
	109	Injured - Misc	29
Total Days			*29*
Squirrel			
	103	Orphan	118
	102	Car Accident	93
Total Days			*211*

Figure 4-7 Report: Final Report (example output)

ASSIGNMENT 3 MAKING A PRESENTATION

Create a presentation that explains the database to Mary Sweeney. Your presentation should be at least 10 minutes in length. Make sure you do the following:

- Explain the design of your database tables.
- Explain how to use the database to run the form, the four queries, and the report.
- Include graphics in your presentation to make it more interesting.
- Describe how your database cuts down on her time for local animal control reporting.
- Discuss future expansion possibilities of the database.

DELIVERABLES

Assemble the following deliverables for your instructor:

1. Word-processed design of tables
2. Tables created in Access
3. Form: One record printed
4. Query 1: Entering Animals
5. Query 2: Length of Stay vs. Type
6. Query 3: Length of Stay vs. Status
7. Query 4: Number of Animals
8. Report: Final Report
9. Presentation graphics, if appropriate
10. Any other required tutorial printouts or tutorial diskette

Staple all pages together. Put your name and class number at the top of the first page. Make sure your diskette is labeled.

5
CASE

The Airline Reservation System

Designing a Relational Database to Create Tables, Forms, Queries, and Reports

Preview

In this case, you will design a relational database for a small start-up airline service. After designing the appropriate tables for the database, you'll populate them with data. Next, you will create a form for entering reservations, two queries for checking flight status, and five reports.

Preparation

- Before attempting this case, you should have experience in database design and in using Microsoft Access.
- Complete any part of Tutorial A that your instructor assigns.
- Complete any part of Tutorial B that your instructor assigns, or refer to the tutorial as necessary.
- Refer to Tutorial E as necessary.

Background

One of your college friends, Frank D'Elia, has decided to start a small airline to fly to selected destinations in the Caribbean. Frank studied business with you in college, and he recently came into an inheritance. He has had a pilot's license since he was a teenager and has always dreamed of owning an airline.

Frank's airline, called "Caribbean Destinations" (CD for short), flies from Philadelphia to the Caribbean and back. The three Caribbean destinations are Grand Bahamas, British Virgin Islands, and Jamaica. Flights each way are scheduled only twice a week. On Mondays and Wednesdays, CD flies to each of the Caribbean destinations from Philadelphia. Then, on Fridays and Sundays, CD flies from the Caribbean back to Philadelphia. Thus, each week there are a total of six departure flights and six return flights. CD needs to keep track of its flights and the passengers who book those flights. CD also needs to know actual take-off and landing times, because they usually differ from the scheduled take-off and landing times.

Although most major airlines have sophisticated decision-support systems to calculate the price of each ticket, CD charges a flat rate for each seat on an aircraft, regardless of the city of origin or the destination. Instead, each ticket's price is a function of when the flight reservation is booked: If a flight is booked 30 days in advance, the price is $100 each way. If a flight is booked 15 days in advance, the price is $150 each way. Any booking made fewer than 15 days in advance costs $200 each way.

CD employs a number of airline pilots and flight attendants. Although each is paid a salary, their hours worked need to be tracked for personnel purposes and for reporting to government agencies. (By law, pilots and flight attendants cannot exceed a certain number of hours of flight time per month.)

Frank recalls some of your database work from college days and asks you to create a database for his company. He would like you to create to a form to be used by his employees who take flight reservations via telephone. He also wants to help those employees answer questions on flight status. You need to create two queries for that—both will be Parameter queries. One query will figure the flight number if the destination city is known. The other query will list the scheduled flight arrival time after a flight number is input.

Frank wants to have some reports created as well. First, he needs to advertise his flights and would like an attractive report for that. Next, because he has to follow strict FAA rules regarding how many hours pilots and flight attendants can work per month, he wants a report that shows employees' names and their hours worked. He would like a summary report, which he will generate every two weeks, showing the previous two weeks of operation. This report should show all flights, their status on each day, and the number of passengers on each flight. In addition, he would like a report that calculates the gross profit for each flight. Finally, if you can think of another report that would be useful for tracking the business, Frank would be very happy to see it.

Assignment 1 Creating the Database Design

In this assignment, you will design your database tables on paper, using a word-processing program. Pay close attention to the tables' logic and structure. Do not start your Access code (Assignment 2) before getting feedback from your instructor on Assignment 1. To design your fields and tables properly, you will need to look at what is required in Assignment 2. Frank has only given you some verbal instructions on this design. Study those closely to come up with the proper tables. When designing the database, observe the following guidelines:

- First, determine the tables you'll need by listing on paper the name of each table and the fields that it should contain. Avoid data redundancy. Do not create a field if it could be created by a "calculated field" in a query.
- You'll need a transaction table. Avoid duplicating data.
- Include a logical field that answers the question, "Landed?"
- Create your tables by using the Table facility of your word processor. Your word-processed tables should resemble the format of the table in Figure 5-1.
- You must mark the appropriate key field(s). You can designate a key field by an asterisk (*) next to the field name. Keep in mind that some tables need a compound primary key.
- Print out the database design.

Table Name	
Field Name	**Data Type (text, numeric, currency, etc.)**
...	...
...	...

Figure 5-1 Table design

Have your design approved before beginning Assignment 2; otherwise, you may need to redo Assignment 2.

ASSIGNMENT 2 CREATING THE DATABASE AND DEVELOPING A FORM, QUERIES, AND REPORTS

Next, you'll create the tables that you'll need. Then you'll create a form, queries, and reports.

Assignment 2A: Creating Tables in Access

In this part of the assignment, you will create your tables in Access. Observe the following guidelines:

- Type records into the tables, using the airport codes of Philadelphia (PHL), Grand Bahamas (FPO), Jamaica (KIN), and British Virgin Islands (VIJ).
- The number of weekly flights are described in the Background section.
- Invent data for two weeks of actual flights (i.e., take-off and landing times). Research commercial flights on the Web to find approximate lengths of flights from Philadelphia to these destinations.
- To minimize typing, assume that there are 36 employees: 18 pilots and 18 flight attendants. Those employees need to be scheduled for flights during a two-week period.

- Use your classmates' names for those of passengers and employees. Type in at least 30 names for passengers. You can duplicate names to minimize typing.
- Appropriately limit the size of the text fields; for example, a flight ID does not need to be the default setting of 50 characters in length.
- Print all tables.

Assignment 2B: Creating a Form, Queries, and Reports

You'll need to create one form, two queries, and five reports, as outlined in the Background section of this case.

Form

CD's reservations personnel need to have a quick way to book passengers on flights. Create a form for this purpose. This form should be based on your transaction table. Print one record from the form.

Query 1 and Query 2

Often, passengers' relatives and friends call the airline to see when a flight is scheduled to land. Create two Parameter queries. Sometimes, people know the flight number of an arriving aircraft, and sometimes they do not. Create one query that calculates which flight number matches which destination. Then make a second query that prompts the user for a flight number and then outputs the resulting scheduled landing time. If a passenger's relative or friend calls with the flight number, the reservation agent can go directly to the second query. Call the first query Find Flight Number on Destin City. Call the second query Find Landing Time on Flight Number.

Report 1

For Frank's advertising purposes, you need to create a report that lists the weekly schedule. Group that report by day of the week. Show the departure, arrival cities, and scheduled times. Make it look attractive enough for advertising in a newspaper. Save the report as Weekly Flights.

Report 2

For recording purposes, Frank needs to know how many hours each pilot and each flight attendant has worked in a two-week period. He will run this report at the end of two-week intervals throughout the year.

To create the report, first create a query that calculates those hours worked, keeping in mind the rules for time arithmetic. Then bring that query into a report and group on "worker." Show the hours by date and then subtotal the hours for each worker, using the Summary Options button in the Report Wizard. Eliminate any extra totals and other extraneous lines generated by the Report Wizard. Save the report as Employees' Hours.

Report 3

There needs to be a summary report that lists all flights in a two-week period, their scheduled arrivals and departures, and their actual arrivals and departures. Your report needs to not only list those times but also count the number of passengers on each flight. You will need to create a query. Make sure the report format looks professional. Call the report Summary Report. You may want to research the FAA rules on the amount of flying both pilots and flight attendants can do during a two-week period. This information may be helpful in describing the usefulness of this report during your presentation.

Report 4

Create a report that will calculate the gross profit for each flight. This needs to be started in a query that calculates the price paid for a ticket in one of the three price categories summed for each of the reservations in that price category. (Remember that price is a function of when a flight was booked.) You will first need to make a query using an IF statement. You may need to ask your instructor for help in forming the IF statement. Then, using that query, create a second query showing flight number, destination and origin cities, date, and profit. Sort in a descending order based on profit so that Frank can see the most profitable flights at the top of the list. Bring this information into the Report Wizard. Make sure that all fields are formatted and that the report looks professional. Call the report Flight Profits.

Report 5

Create another report that you think would be useful for this start-up airline. It can be based on a table or a query. On the bottom of the report, hand-write a note that describes the question or business problem that this report addresses. Give the report a descriptive title and save it.

Assignment 3 Making a Presentation

Create a presentation that explains the database to Frank and all the reservations agents. Your presentation should take at least 20 minutes. Make sure you do the following:

- Explain the design of your database tables.
- Expain how to use the database and run the form, queries, and reports.
- Include some graphics in your presentation to make it more interesting.
- Discuss the rationale behind Report 5, the report that you designed.
- Discuss how you could migrate this project to the Web.
- Show a mock-up of what a future Web site could look like, including graphics and links.

Deliverables

Assemble the following deliverables for your instructor:

1. Word-processed design of tables
2. Tables created in Access
3. Form: One record printed
4. Query 1: Find Flight Number on Destin City
5. Query 2: Find Landing Time on Flight Number
6. Report 1: Weekly Flights
7. Report 2: Employees' Hours
8. Report 3: Summary Report
9. Report 4: Flight Profits
10. Report 5: (your own report)
11. Presentation graphics or handouts, as appropriate
12. Any other required tutorial printouts or tutorial diskette

Staple all pages together. Put your name and class number at the top of the first page. Make sure your diskette is labeled.

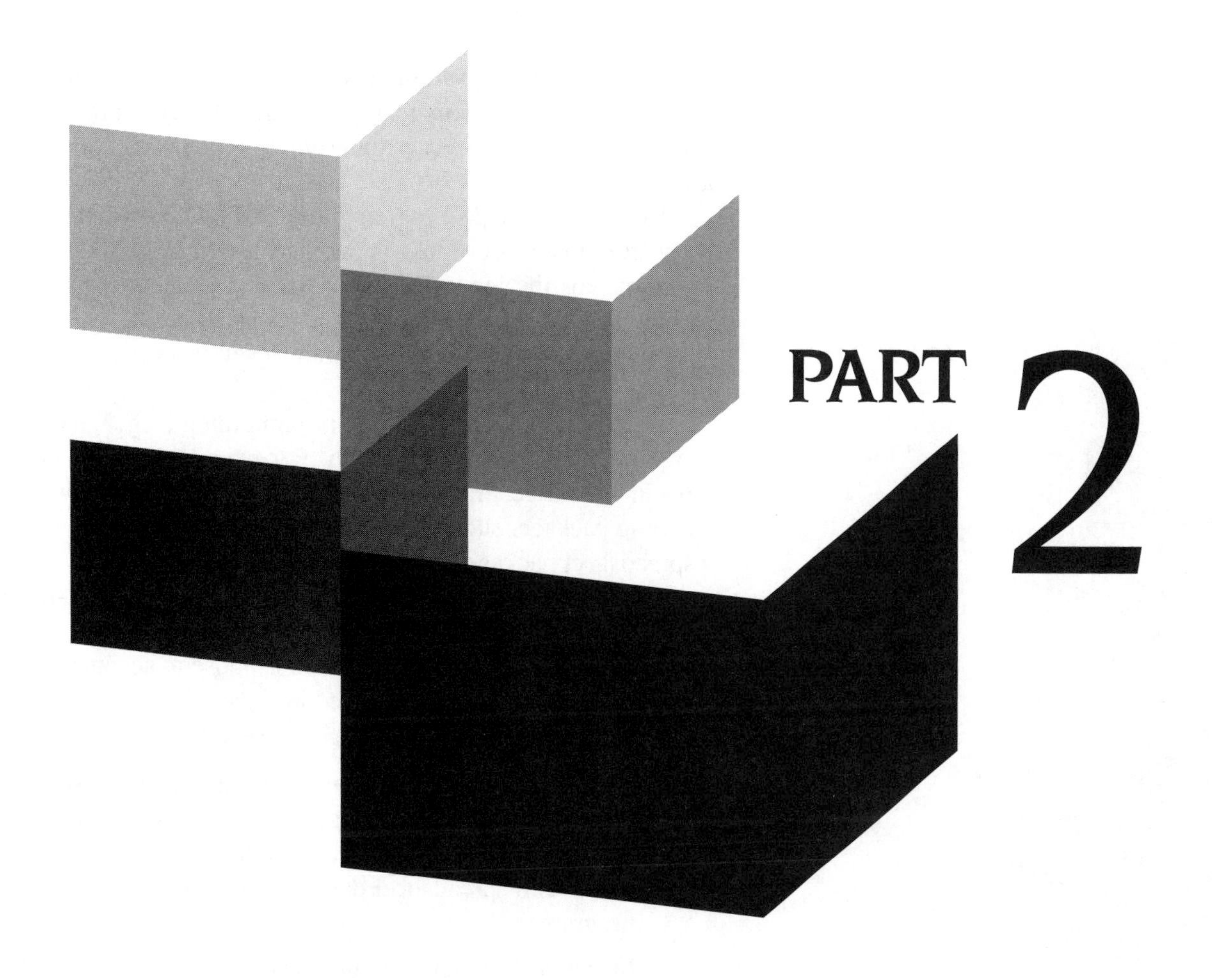

PART 2

Decision Support Cases Using Excel Scenario Manager

C
TUTORIAL

Building a Decision Support System in Excel

A **decision support system (DSS)** is a computer program that can represent, either mathematically or symbolically, a problem that a user needs to solve. Such a representation is, in effect, a model of a problem.

Here's how a DSS program works: The DSS program accepts input from the user or looks at data in files on disk. Then, the DSS program runs the input and any other necessary data through the model. The program's output is the information the user needs to solve a problem. Some DSS programs even recommend a solution to a problem.

A DSS can be written in any programming language that lets a programmer represent a problem. For example, a DSS could be built in a third-generation language, such as Visual Basic, or in a database package, such as Access. A DSS could also be written in a spreadsheet package, such as Excel.

The Excel spreadsheet package has standard built-in arithmetic functions as well as many statistical and financial functions. Thus, many kinds of problems—such as those in accounting, operations, or finance—can be modeled in Excel.

This tutorial has the following four sections:

1. **Spreadsheet and DSS Basics:** In this section, you'll learn how to create a DSS program in Excel. Your program will be in the form of a cash flow model. This will give you practice in spreadsheet design and in building a DSS program.
2. **Scenario Manager:** In this section, you'll learn how to use an Excel tool called the Scenario Manager. With any DSS package, one problem with playing "what if" is this: Where do you physically record the results from running each set of data? Typically, a user just writes the inputs and related results on a piece of paper. Then—ridiculously enough—the user might have to input the data *back* into a spreadsheet for further analysis! The Scenario Manager solves that problem. It can be set up to capture inputs and results as "scenarios," which are then nicely summarized on a separate sheet in the Excel workbook.
3. **Practice Using Scenario Manager:** You are invited to work on a different problem, a case using the Scenario Manager.
4. **Review of Excel Basics:** This section reviews the information you'll need to do the spreadsheet cases that follow this tutorial.

SPREADSHEET AND DSS BASICS

Assume it is late in 2004, and that you are trying to build a model of what a company's net income (profit) and cash flow will be in the next two years (2005 and 2006). This is the problem: to forecast net income and cash flow in those years. Assume that knowing these forecasts would help to answer some question or make some decision. After researching the problem, you decide that the estimates should be based on three things: (1) 2004 results, (2) estimates of the underlying economy, and (3) the cost of products the company sells.

The model will use an income statement and cash flow framework. The user can input values for two possible states of the economy in years 2005-2006: an "O" for an Optimistic outlook and a "P" for a Pessimistic outlook. The state of the economy is expected to affect the number of units the company can sell as well as the unit selling price: In a good "O" economy, more units can be sold at a higher price. The user can also input values for two possible cost-of-goods-sold price directions: a "U" for Up or a "D" for Down. A "U" means that the cost of an item sold will be higher than in 2004; a "D" means that it will be less.

Presumably, the company will do better in a good economy and with lower input costs—but how much better? The relationships are too complex to assess in one's head, but the software model can easily assess the relationships. Thus, the user can play "what if" with the input variables and note the effect on net income and year-end cash levels. For example, a user can ask, "What if the economy is good and costs go up? What will net income and cash flow be in that case? What would happen if the economy is down and costs go down? What would be the company's net income and cash flow in that case?" With an Excel software model available, the answers are easily quantified.

Organization of the DSS Model

Your spreadsheets should have the following sections, which will be noted in boldface type throughout this tutorial and in the Excel cases that follow it:

- **CONSTANTS**
- **INPUTS**
- **SUMMARY OF KEY RESULTS**
- **CALCULATIONS** (of values that will be used in the INCOME STATEMENT AND CASH FLOW STATEMENT)
- **INCOME STATEMENT AND CASH FLOW STATEMENT**

Here, as an extended illustration, a DSS model is built for the forecasting problem previously described. Let's look at each spreadsheet section. Figures C-1 and C-2 show how to set up the spreadsheet.

	A	B	C	D
1	**TUTORIAL EXERCISE**			
2				
3	**CONSTANTS**	**2004**	**2005**	**2006**
4	TAX RATE	NA	0.33	0.35
5	NUMBER OF BUSINESS DAYS	NA	300	300
6				
7	**INPUTS**	**2004**	**2005**	**2006**
8	ECONOMIC OUTLOOK (O=OPTIMISTIC; P=PESSIMISTIC)	NA		NA
9	PURCHASE-PRICE OUTLOOK (U=UP; D=DOWN)	NA		NA
10				
11	**SUMMARY OF KEY RESULTS**	**2004**	**2005**	**2006**
12	NET INCOME AFTER TAXES	NA		
13	END-OF-THE-YEAR CASH ON HAND	NA		
14				
15	**CALCULATIONS**	**2004**	**2005**	**2006**
16	NUMBER OF UNITS SOLD IN A DAY	1000		
17	SELLING PRICE PER UNIT	7.00		
18	COST OF GOODS SOLD PER UNIT	3.00		
19	NUMBER OF UNITS SOLD IN A YEAR	NA		

Figure C-1 Tutorial skeleton 1

	A	B	C	D
21	**INCOME STATEMENT AND CASH FLOW STATEMENT**	**2004**	**2005**	**2006**
22	BEGINNING-OF-THE-YEAR CASH ON HAND	NA		
23				
24	SALES (REVENUE)	NA		
25	COST OF GOODS SOLD	NA		
26	INCOME BEFORE TAXES	NA		
27	INCOME TAX EXPENSE	NA		
28	NET INCOME AFTER TAXES	NA		
29				
30	END-OF-THE-YEAR CASH ON HAND (BEGINNING-OF-THE-YEAR CASH PLUS NET INCOME AFTER TAXES)	10000		

Figure C-2 Tutorial skeleton 2

The CONSTANTS Section

This section records values that are used in spreadsheet calculations. In a sense, the constants are inputs, except that they do not change. In this tutorial, constants are TAX RATE and the NUMBER OF BUSINESS DAYS.

The INPUTS Section

The inputs are for the ECONOMIC OUTLOOK and PURCHASE-PRICE OUTLOOK (manufacturing input costs). Inputs could conceivably be entered for *each year* covered by the model (here, 2005 and 2006). This would let you enter an "O" for 2005's economy in one cell and a "P" for 2006's economy in another cell. Alternately, one input for the two-year period could be entered in one cell. For simplicity, this tutorial uses the *latter* approach.

The Summary of Key Results Section

This section will capture 2005 and 2006 NET INCOME AFTER TAXES (profit) for the year and END-OF-THE-YEAR CASH ON HAND, which are (assume) the two relevant outputs of this model. The summary merely repeats, in one easy-to-see place, results that are shown in otherwise widely spaced places in the spreadsheet. This just makes the answers easier to see all at once. (It also makes it easier to graph results later.)

The CALCULATIONS Section

This area is used to compute the following data:

1. The NUMBER OF UNITS SOLD IN A DAY (a function of a certain 2004 value and of the input economic outlook)
2. The SELLING PRICE PER UNIT (similarly derived)
3. The COST OF GOODS SOLD PER UNIT (a function of a 2004 value and of the purchase-price outlook)
4. The NUMBER OF UNITS SOLD IN A YEAR (the number of units sold in a day times the number of business days)

These formulas could be embedded in the **INCOME STATEMENT AND CASH FLOW STATEMENT** section. Doing that would, however, cause the expressions there to be complex and difficult to understand. Putting the intermediate calculations into a separate **CALCULATIONS** section breaks up the work into modules. This is good form because it simplifies your programming.

The INCOME STATEMENT and CASH FLOW STATEMENT Section

This is the "body" of the spreadsheet. It shows the following:

1. BEGINNING-OF-THE-YEAR CASH ON HAND, which equals cash at the end of the *prior* year.
2. SALES (REVENUE), which equals the units sold in the year times the unit selling price.
3. COST OF GOODS SOLD, which is units sold in the year times the price paid to acquire or make the unit sold.
4. INCOME BEFORE TAXES, which equals sales less total costs.
5. INCOME TAX EXPENSE, which is zero if there are losses; otherwise, it is the pre-tax margin times the constant tax rate. (This is sometimes called INCOME TAXES.)
6. NET INCOME AFTER TAXES, which equals income before taxes less income tax expense.
7. END-OF-THE-YEAR CASH ON HAND is beginning-of-the-year cash on hand plus net income. (In the real world, cash flow estimates must account for changes in receivables and payables. In this case, assume that sales are collected immediately—i.e., there are no receivables or bad debts. Assume also that suppliers are paid immediately—i.e., that there are no payables.)

Construction of the Spreadsheet Model

Next, let's work through the following three steps to build your spreadsheet model:

1. Make a "skeleton" of the spreadsheet, and call it **TUTC.xls**.
2. Fill in the "easy" cell formulas.
3. Enter the "hard" spreadsheet formulas.

Make a Skeleton

Your first step is to set up a skeleton worksheet. This should have headings, text string labels, and constants—but no formulas.

To set up the skeleton, you must get a grip on the problem, *conceptually*. The best way to do that is to work *backward* from what the "body" of the spreadsheet will look like. Here, the body is the **INCOME STATEMENT AND CASH FLOW STATEMENT** section. Set that up, in your mind or on paper, then do the following:

- Decide what amounts should be in the **CALCULATIONS** section. In this tutorial's model, SALES (revenue) will be NUMBER OF UNITS SOLD IN A DAY times SELLING PRICE PER UNIT, in the income statement. You will calculate the intermediate amounts (NUMBER OF UNITS SOLD IN A YEAR and SELLING PRICE PER UNIT) in the **CALCULATIONS** section.
- Set up the **SUMMARY OF KEY RESULTS** section by deciding what *outputs* are needed to solve the problem. The **INPUTS** section should be reserved for amounts that can change—the controlling variables—which are the ECONOMIC OUTLOOK and the PURCHASE-PRICE OUTLOOK.
- Use the **CONSTANTS** section for values that you will need to use, but that are not in doubt, i.e., you will not have to input them or calculate them. Here, the TAX RATE is a good example of such a value.

AT THE KEYBOARD

Type in the Excel skeleton shown in Figures C-1 and C-2.

A designation of "NA" means that a cell will not be used in any formula in the worksheet. The 2004 values are needed only for certain calculations, so for the most part, the 2004 column's cells just show "NA." (Recall that the forecast is for 2005 and 2006.) Also be aware that you can "break" a text string in a cell by pressing the keys Alt and Enter at the same time at the break point. This makes the cell "taller." Formatting of cells to show centered data and creation of borders is discussed at the end of this tutorial.

Fill in the "Easy" Formulas

The next step is to fill in the "easy" formulas. The cells affected (and what you should enter) are discussed next.

To prepare, you should format the cells in the **SUMMARY OF KEY RESULTS** section for no decimals. (Formatting for numerical precision is discussed at the end of this tutorial.) The **SUMMARY OF KEY RESULTS** section just "echoes" results shown in other places. The idea is that C28 holds the NET INCOME AFTER TAXES. You want to echo that amount in C12. So, the formula in C12 is =C28. Translation: "Copy what is in C28 into C12." It's that simple.

With the insertion point in C12, the contents of that cell—in this case the formula =C28—shows in the editing window, which is above the lettered column indicators, as shown in Figure C-3.

C12	fx =C28		
	A	B	C	D
11	**SUMMARY OF KEY RESULTS**	2004	2005	2006
12	NET INCOME AFTER TAXES	NA	0	
13	END-OF-THE-YEAR CASH ON HAND	NA		

Figure C-3 Echo 2005 NET INCOME AFTER TAXES

At this point, C28 is empty (and thus has a zero value), but that does not prevent you from copying. So, copy cell C12's formula to the right, to cell D12. The copy operation does not actually "copy." Copying puts =D28 into D12, which is what you want. (Year 2006's NET INCOME AFTER TAXES is in D28.)

To perform the Copy operation, use the following steps:

1. Select (click in) the cell (or range of cells) that you want to copy.
2. Choose Edit—Copy.
3. Select the cell (or cell range) to be copied to by clicking (and then dragging if the range has more than one cell).
4. Press the Enter key.

END-OF-THE-YEAR CASH ON HAND for 2005 cash is in cell C30. Echo the cash results in cell C30 to cell C13. (Put =C30 in cell C13 as shown in Figure C-4.) Copy the formula from C13 to D13.

C13	fx =C30		
	A	B	C	D
11	**SUMMARY OF KEY RESULTS**	2004	2005	2006
12	NET INCOME AFTER TAXES	NA	0	0
13	END-OF-THE-YEAR CASH ON HAND	NA	0	

Figure C-4 Echo 2005 END-OF-THE-YEAR CASH ON HAND

At this point, the **CALCULATIONS** section formulas will not be entered because they are not all "easy" formulas. Move on to the easier formulas in the **INCOME STATEMENT AND CASH FLOW STATEMENT** section, as if the calculations were already done. Again, the empty **CALCULATIONS** section cells in this section do not stop you from entering formulas. You should now format the cells in the **INCOME STATEMENT AND CASH FLOW STATEMENT** section for zero decimals.

BEGINNING-OF-THE-YEAR CASH ON HAND is the cash on hand at the end of the *prior* year. In C22 for the year 2005, type =B30. See the "skeleton" you just entered, as shown in Figure C-5. Cell B30 has the END-OF-THE-YEAR CASH ON HAND for 2004. Copy the formula to the right.

C22	fx =B30		
	A	**B**	**C**	**D**

	A	B	C	D
21	**INCOME STATEMENT AND CASH FLOW STATEMENT**	**2004**	**2005**	**2006**
22	BEGINNING-OF-THE-YEAR CASH ON HAND	NA	10000	
23				
24	SALES (REVENUE)	NA		
25	COST OF GOODS SOLD	NA		
26	INCOME BEFORE TAXES	NA		
27	INCOME TAX EXPENSE	NA		
28	NET INCOME AFTER TAXES	NA		
29				
30	END-OF-THE-YEAR CASH ON HAND (BEGINNING-OF-THE-YEAR CASH PLUS NET INCOME AFTER TAXES)	10000		

Figure C-5 Echo of END-OF-THE-YEAR CASH ON HAND for 2004 to BEGINNING-OF-THE-YEAR CASH ON HAND for 2005

Sales (REVENUE) is just NUMBER OF UNITS SOLD IN A YEAR times SELLING PRICE PER UNIT. In cell C24 enter =C17*C19, as shown in Figure C-6.

C24 fx =C17*C19

	A	B	C	D
15	**CALCULATIONS**	**2004**	**2005**	**2006**
16	NUMBER OF UNITS SOLD IN A DAY	1000		
17	SELLING PRICE PER UNIT	7.00		
18	COST OF GOODS SOLD PER UNIT	3.00		
19	NUMBER OF UNITS SOLD IN A YEAR	NA		
20				
21	**INCOME STATEMENT AND CASH FLOW STATEMENT**	**2004**	**2005**	**2006**
22	BEGINNING-OF-THE-YEAR CASH ON HAND	NA	10000	0
23				
24	SALES (REVENUE)	NA	0	
25	COST OF GOODS SOLD	NA		

Figure C-6 Enter the formula to compute 2005 SALES

The formula C17*C19 multiplies units sold for the year times the unit selling price. (Cells C17 and C19 are empty now, which is why SALES shows as zero after the formula is entered.) Copy the formula to the right, to D24.

COST OF GOODS SOLD is handled similarly. In C25, type =C18*C19. This equals NUMBER OF UNITS SOLD IN A YEAR times COST OF GOODS SOLD PER UNIT. Copy to the right.

In cell C26, the formula for INCOME BEFORE TAXES is =C24–C25. Enter the formula. Copy to the right.

In the United States, income taxes are only paid on positive income before taxes. In cell C27, the INCOME TAX EXPENSE is zero if the INCOME BEFORE TAXES is zero or less; else, INCOME TAX EXPENSE equals the TAX RATE times the INCOME BEFORE TAXES. The TAX RATE is a constant (in C4). An =IF() statement is needed to express this logic:

IF(INCOME BEFORE TAXES is <= 0, put zero tax in C27,
else in C27 put a number equal to multiplying the
TAX RATE times the INCOME BEFORE TAXES)

C26 stands for the concept INCOME BEFORE TAXES, and C4 stands for the concept TAX RATE. So, in Excel, substitute those cell addresses:

=IF(C26 <= 0, 0, C4 * C26)

Copy the income tax expense formula to the right.

In cell C28, NET INCOME AFTER TAXES is just INCOME BEFORE TAXES less INCOME TAX EXPENSE: =C26-C27. Enter and copy to the right.

The END-OF-THE-YEAR CASH ON HAND is BEGINNING-OF-THE-YEAR CASH ON HAND plus NET INCOME AFTER TAXES. In cell C30, enter =C22+C28. The **INCOME STATEMENT AND CASH FLOW STATEMENT** section at that point is shown in Figure C-7. Then, copy the formula to the right.

Tutorial C

C30 fx =C22+C28

	A	B	C	D
21	**INCOME STATEMENT AND CASH FLOW STATEMENT**	**2004**	**2005**	**2006**
22	BEGINNING-OF-THE-YEAR CASH ON HAND	NA	10000	10000
23				
24	SALES (REVENUE)	NA	0	0
25	COST OF GOODS SOLD	NA	0	0
26	INCOME BEFORE TAXES	NA	0	0
27	INCOME TAX EXPENSE	NA	0	0
28	NET INCOME AFTER TAXES	NA	0	0
29				
30	END-OF-THE-YEAR CASH ON HAND (BEGINNING-OF-THE-YEAR CASH PLUS NET INCOME AFTER TAXES)	10000	10000	

Figure C-7 Status of INCOME STATEMENT AND CASH FLOW STATEMENT

Put in the "Hard" Formulas

The next step is to finish the spreadsheet by filling in the "hard" formulas.

AT THE KEYBOARD

First, in C8 enter an "O" (no quotation marks) for OPTIMISTIC, and in C9 enter "U" (no quotation marks) for UP. There is nothing magic about these particular values—they just give the worksheet formulas some input to process. Recall that the inputs will cover both 2005 and 2006. Make sure "NA" is in D8 and D9, just to remind yourself that these cells will not be used for input or by other worksheet formulas. Your **INPUTS** section should look like the one shown in Figure C-8.

	A	B	C	D
7	**INPUTS**	**2004**	**2005**	**2006**
8	ECONOMIC OUTLOOK (O=OPTIMISTIC; P=PESSIMISTIC)	NA	O	NA
9	PURCHASE-PRICE OUTLOOK (U=UP; D=DOWN)	NA	U	NA

Figure C-8 Entering two input values

Recall that cell addresses in the **CALCULATIONS** section are already referred to in formulas in the **INCOME STATEMENT AND CASH FLOW STATEMENT** section. The next step is to enter formulas for these calculations. Before doing that, format NUMBER OF UNITS SOLD IN A DAY and NUMBER OF UNITS SOLD IN A YEAR for zero decimals, and format SELLING PRICE PER UNIT and COST OF GOODS SOLD PER UNIT for two decimals.

The easiest formula in the **CALCULATIONS** section is the NUMBER OF UNITS SOLD IN A YEAR, which is the calculated NUMBER OF UNITS SOLD IN A DAY (in C16) times the NUMBER OF BUSINESS DAYS (in C5). In C19, enter =C5*C16, as shown in Figure C-9.

C19 fx =C5*C16

	A	B	C	D
1	**TUTORIAL EXERCISE**			
2				
3	**CONSTANTS**	**2004**	**2005**	**2006**
4	TAX RATE	NA	0.33	0.35
5	NUMBER OF BUSINESS DAYS	NA	300	300
6				
7	**INPUTS**	**2004**	**2005**	**2006**
8	ECONOMIC OUTLOOK (O=OPTIMISTIC; P=PESSIMISTIC)	NA	O	NA
9	PURCHASE-PRICE OUTLOOK (U=UP; D=DOWN)	NA	U	NA
10				
11	**SUMMARY OF KEY RESULTS**	**2004**	**2005**	**2006**
12	NET INCOME AFTER TAXES	NA	0	0
13	END-OF-THE-YEAR CASH ON HAND	NA	10000	10000
14				
15	**CALCULATIONS**	**2004**	**2005**	**2006**
16	NUMBER OF UNITS SOLD IN A DAY	1000		
17	SELLING PRICE PER UNIT	7.00		
18	COST OF GOODS SOLD PER UNIT	3.00		
19	NUMBER OF UNITS SOLD IN A YEAR	NA	0	

Figure C-9 Entering the formula to compute 2005 NUMBER OF UNITS SOLD IN A YEAR

Copy the formula to cell D19, for year 2006.

Assume that if the ECONOMIC OUTLOOK is OPTIMISTIC, the 2005 NUMBER OF UNITS SOLD IN A DAY will be 6% more than that in 2004; in 2006, they will be 6% more than that in 2005. Also assume that if the ECONOMIC OUTLOOK is PESSIMISTIC, the NUMBER OF UNITS SOLD IN A DAY in 2005 will be 1% less than those sold in 2004; in

2006, they will be 1% less than those sold in 2005. An =IF() statement is needed in C16 to express this idea:

IF(economy variable = OPTIMISTIC,
then NUMBER OF UNITS SOLD IN A DAY will go UP 6%,
else NUMBER OF UNITS SOLD IN A DAY will go DOWN 1%)

Substituting cell addresses:

=IF(C8 = "O", 1.06 * B16, .99 * B16)

In Excel, quotation marks denote labels. The input is a one-letter label. So, the quotation marks around the **'O'** are needed. You should also note that multiplying by 1.06 results in a 6% rise, whereas multiplying by .99 results in a 1% decrease.

Enter the entire =IF formula into cell C16, as shown in Figure C-10. Absolute addressing is needed (C8) because the address is in a formula that gets copied, *and* you do not want this cell reference to change (to D8, which has the value "NA") when you copy the formula to the right. Absolute addressing maintains the C8 reference when the formula is copied. Copy the formula in C16 to D16 for 2006.

C16 =IF(C8="O",1.06*B16,0.99*B16)

	A	B	C	D
15	**CALCULATIONS**	**2004**	**2005**	**2006**
16	NUMBER OF UNITS SOLD IN A DAY	1000	1060	
17	SELLING PRICE PER UNIT	7.00		
18	COST OF GOODS SOLD PER UNIT	3.00		
19	NUMBER OF UNITS SOLD IN A YEAR	NA	318000	0

Figure C-10 Entering the formula to compute 2005 NUMBER OF UNITS SOLD IN A DAY

The SELLING PRICE PER UNIT is also a function of the ECONOMIC OUTLOOK. The two-part rule is (assume) as follows:

- If the ECONOMIC OUTLOOK is OPTIMISTIC, the SELLING PRICE PER UNIT in 2005 will be 1.07 times that of 2004; in 2006 it will be 1.07 times that of 2005.
- On the other hand, if the ECONOMIC OUTLOOK is PESSIMISTIC, the SELLING PRICE PER UNIT in 2005 and 2006 will equal the per-unit price in 2004 (that is, the price will not change).

Test your understanding of the selling price calculation by figuring out what the formula should be for cell C17. Enter it and copy to the right. You will need to use absolute addressing. (Can you see why?)

The COST OF GOODS SOLD PER UNIT is a function of the PURCHASE-PRICE OUTLOOK. The two-part rule is (assume) as follows:

- If the PURCHASE-PRICE OUTLOOK is UP ("U"), COST OF GOODS SOLD PER UNIT in 2005 will be 1.25 times that of year 2004; in 2006, it will be 1.25 times that of 2005.
- On the other hand, if the PURCHASE-PRICE OUTLOOK is DOWN ("D"), the multiplier in each year will be 1.01.

Again, to test your understanding, figure out what the formula should be in cell C18. Enter it and copy to the right. You will need to use absolute addressing.

Your selling price and cost of goods sold formulas, given OPTIMISTIC and UP input values, should yield the calculated values shown in Figure C-11.

	A	B	C	D
15	**CALCULATIONS**	**2004**	**2005**	**2006**
16	NUMBER OF UNITS SOLD IN A DAY	1000	1060	1124
17	SELLING PRICE PER UNIT	7.00	7.49	8.01
18	COST OF GOODS SOLD PER UNIT	3.00	3.75	4.69
19	NUMBER OF UNITS SOLD IN A YEAR	NA	318000	337080

Figure C-11 Calculated values given OPTIMISTIC and UP input values

Assume that you change the input values to PESSIMISTIC and DOWN. Your formulas should yield the calculated values shown in Figure C-12.

	A	B	C	D
15	**CALCULATIONS**	**2004**	**2005**	**2006**
16	NUMBER OF UNITS SOLD IN A DAY	1000	990	980
17	SELLING PRICE PER UNIT	7.00	7.00	7.00
18	COST OF GOODS SOLD PER UNIT	3.00	3.03	3.06
19	NUMBER OF UNITS SOLD IN A YEAR	NA	297000	294030

Figure C-12 Calculated values given PESSIMISTIC and DOWN input values

That completes the body of your spreadsheet! The values in the **CALCULATIONS** section ripple through the **INCOME STATEMENT AND CASH FLOW STATEMENT** section because the income statement formulas reference the calculations. Assuming inputs of OPTIMISTIC and UP, the income and cash flow numbers should now look like those in Figure C-13.

	A	B	C	D
21	**INCOME STATEMENT AND CASH FLOW STATEMENT**	**2004**	**2005**	**2006**
22	BEGINNING-OF-THE-YEAR CASH ON HAND	NA	10000	806844
23				
24	SALES (REVENUE)	NA	2381820	2701460
25	COST OF GOODS SOLD	NA	1192500	1580063
26	INCOME BEFORE TAXES	NA	1189320	1121398
27	INCOME TAX EXPENSE	NA	392476	392489
28	NET INCOME AFTER TAXES	NA	796844	728909
29				
30	END-OF-THE-YEAR CASH ON HAND (BEGINNING-OF-THE-YEAR CASH PLUS NET INCOME AFTER TAXES)	10000	806844	1535753

Figure C-13 Completed INCOME STATEMENT AND CASH FLOW STATEMENT section

SCENARIO MANAGER

You are now ready to use the Scenario Manager to capture inputs and results as you play "what if" with the spreadsheet.

Note that there are four possible combinations of input values: O-U (Optimistic-Up), O-D (Optimistic-Down), P-U (Pessimistic-Up), and P-D (Pessimistic-Down). Financial results for each combination will be different. Each combination of input values can be referred to as a "scenario." Excel's Scenario Manager records the results of each combination of input values as a separate scenario and then shows a summary of all scenarios in a separate worksheet. These summary worksheet values can be used as a raw table of numbers, which could be printed or copied into a Word document. The table of data could then be the basis for an Excel chart, which could also be printed or put into a memorandum.

You have four possible scenarios for the economy and the purchase price of goods sold: Optimistic-Up, Optimistic-Down, Pessimistic-Up, and Pessimistic-Down. The four input-value sets produce different financial results. When you use the Scenario Manager, define the four scenarios, then have Excel (1) sequentially run the input values "behind the scenes," and then (2) put the results for each input scenario in a summary sheet.

When you define a scenario, you give the scenario a name and identify the input cells and input values. You do this for each scenario. Then you identify the output cells, so Excel can capture the outputs in a Summary Sheet.

Tutorial C

AT THE KEYBOARD

To start, select Tools—Scenarios. This leads you to a Scenario Manager window. Initially, there are no scenarios defined, and Excel tells you that, as you can see in Figure C-14.

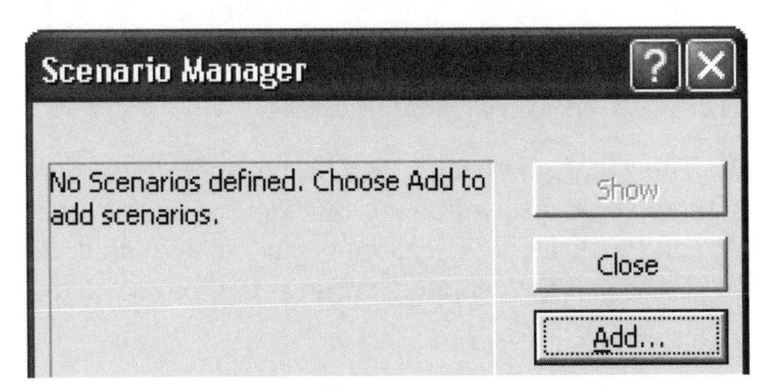

Figure C-14 Initial Scenario Manager window

With this window, you are able to add scenarios, delete them, or change (edit) them. Toward the end of the process, you are also able to create the summary sheet.

NOTE

When working with this window and its successors, do **not** hit the Enter key to navigate. Use mouse clicks to move from one step to the next.

To continue with defining a scenario: Click the Add button. In the resulting Add Scenario window, give the first scenario a name: OPT-UP. Then type in the input cells in the Changing cells window, here, C8:C9. (*Note*: C8 and C9 are contiguous input cells. Non-contiguous input cell ranges can be separated by a comma.) Excel may add dollar signs to the cell address—do not be concerned about this. The window should look like the one shown in Figure C-15.

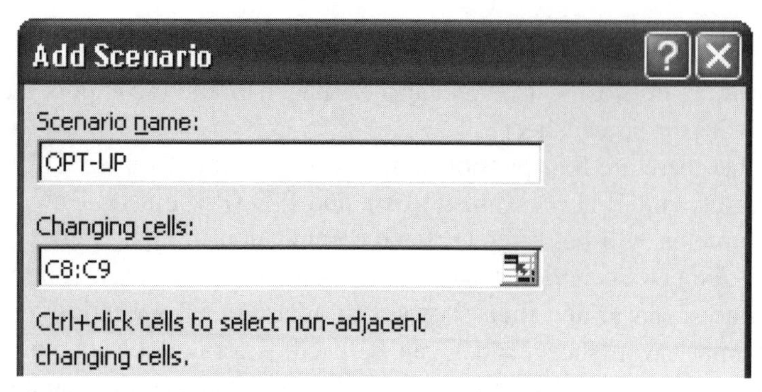

Figure C-15 Entering OPT-UP as a scenario

Now click OK. This moves you to the Scenario Values window. Here you indicate what the INPUT *values* will be for the scenario. The values in the cells *currently in* the spreadsheet will be displayed. They might—or might not—be correct for the scenario you are defining. For the OPT-UP scenario, an O and a U would need to be entered, if not the current values. Enter those values if need be, as shown in Figure C-16.

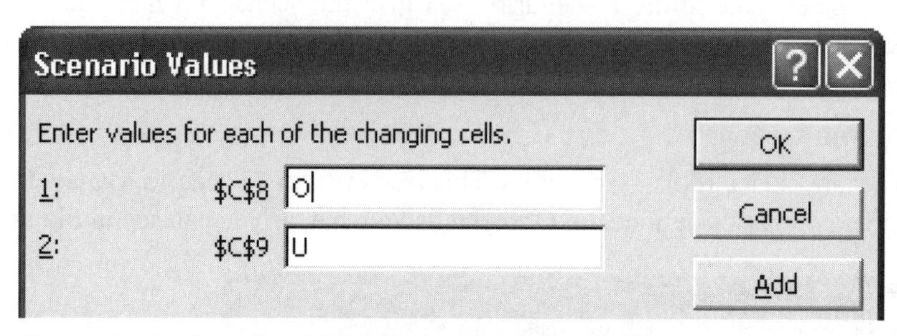

Figure C-16 Entering OPT-UP scenario input values

Now click OK. This takes you back to the Scenario Manager window. You are now able to enter the other three scenarios, following the same steps. Do that now! Enter the OPT-DOWN, PESS-UP, and PESS-DOWN scenarios, plus related input values. After all that, you should see that the names and the changing cells for the four scenarios have been entered, as in Figure C-17.

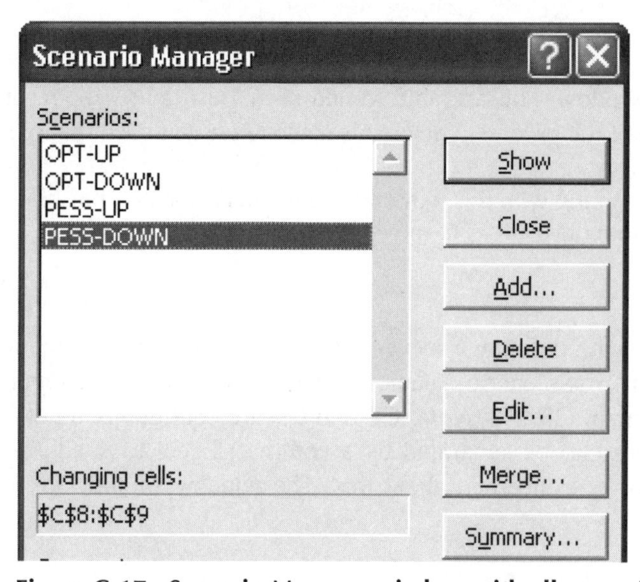

Figure C-17 Scenario Manager window with all scenarios entered

You are now able to create a summary sheet that shows the results of running the four scenarios. Click the Summary button. You'll get the Scenario Summary window. You must tell Excel what the output cell addresses are—these will be the same for all four scenarios. (The output *values* change in those output cells as input values are changed, but the addresses of the output cells do not change.)

Assume that you are primarily interested in the results that have accrued at the end of the two-year period. These are your two 2006 **SUMMARY OF KEY RESULTS** section cells for NET INCOME AFTER TAXES and END-OF-THE-YEAR CASH ON HAND (D12 and D13). Type these addresses into the window's input area, as shown in Figure C-18. (*Note*: If results cells are non-contiguous, the address ranges can be entered, separated by a comma.)

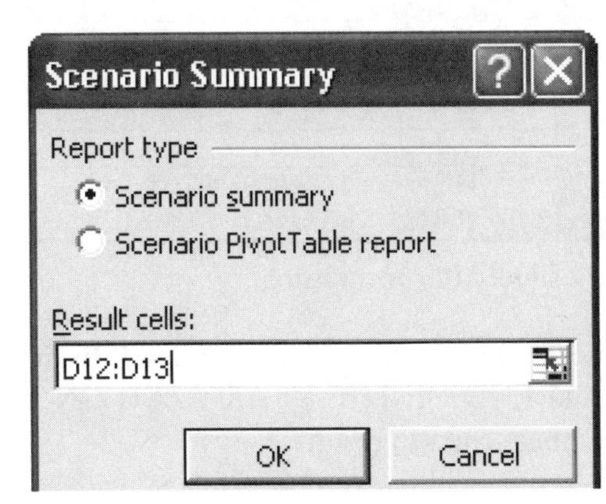

Figure C-18 Entering Result Cells addresses in Scenario Summary window

Then click OK. Excel runs each set of inputs and collects results as it goes. (You do not see this happening on the screen.) Excel makes a *new* sheet, titled the Scenario Summary (denoted by the sheet's lower tab), and takes you there, as shown in Figure C-19.

Scenario Summary	Current Values:	OPT-UP	OPT-DOWN	PESS-UP	PESS-DOWN
Changing Cells:					
C8	O	O	O	P	P
C9	U	U	D	U	D
Result Cells:					
D12	728909	728909	1085431	441964	752953
D13	1535753	1535753	2045679	1098681	1552944

Notes: Current Values column represents values of changing cells at time Scenario Summary Report was created. Changing cells for each scenario are highlighted in gray.

Figure C-19 Scenario Summary sheet created by Scenario Manager

One somewhat annoying visual element is that the Current Values in the spreadsheet itself are given an output column. This duplicates one of the four defined scenarios. You can delete that extra column by (1) clicking on its column designator letter (here, column D), and (2) clicking Edit—Delete.

Do *not* select Edit—Delete *Sheet*!

Another annoyance is that Column A goes unused. You can click and delete it in the same way to move everything over to the left. This should make columns of data easier to see on the screen, without scrolling. You can also (1) edit cell values to make the results more clear, (2) enter words for the cell addresses, (3) use Alt—Enter to break long headings, if need be, (4) center values using the Format—Cells—Alignment tab menu option, and (5) show data in Currency format, Using the Format—Cells—Number tab menu option.

When you're done, your summary sheet should resemble the one shown in Figure C-20.

	A	B	C	D	E	F
1	**Scenario Summary**		OPTIMISTIC	OPTIMISTIC	PESSIMISTIC	PESSIMISTIC
2			UP	DOWN	UP	DOWN
4	**Changing Cells:**					
5	**ECONOMIC OUTLOOK**	**C8**	O	O	P	P
6	**PURCHASE PRICE OUTLOOK**	**C9**	U	D	U	D
7	**Result Cells (2005):**					
8	**NET INCOME AFTER TAXES**	**D12**	$728,909	$1,085,431	$441,964	$752,953
9	**END-OF-THE-YEAR CASH ON HAND**	**D13**	$1,535,753	$2,045,679	$1,098,681	$1,552,944

Figure C-20 Scenario Summary sheet after formatting

Note that column C shows the OPTIMISTIC-UP case. NET INCOME AFTER TAXES in that scenario is $728,909, and END-OF-THE-YEAR CASH ON HAND is $1,535,753. Columns D, E, and F show the other scenario results.

Here is an important post-script to this exercise: DSS spreadsheets are used to guide decision-making. This means that the spreadsheet's results must be interpreted in some way. Let's practice with the results shown in Figure C-20. With that data, what combination of year 2006 NET INCOME AFTER TAXES and END-OF-THE-YEAR CASH ON HAND would be best?

Clearly, O-D is the best result, right? It yields the highest income and highest cash. What is the worst combination? P-U, right? It yields the lowest income and lowest cash.

Results are not always this easy to interpret, but the analytical method is the same. You have a complex situation that you cannot understand very well without software assistance. You build a model of the situation in the spreadsheet, enter the inputs, collect the results, and then interpret the results to guide decision-making.

Summary Sheets

When you do Scenario Manager spreadsheet case studies, you'll need to manipulate Summary Sheets and their data. Let's look at some of these operations.

Rerunning the Scenario Manager

To rerun the Scenario Manager, click the Summary button in the Scenario Manager dialog box and then click the OK button. This makes another summary sheet. It does not overwrite a prior one.

Deleting Unwanted Scenario Manager Summary Sheets

Suppose that you want to delete a Summary sheet. With the Summary sheet on the screen, select Edit—Delete Sheet. You will be asked if you really mean it. If so, click to remove, or else cancel out.

Charting Summary Sheet Data

The Summary sheet results can be conveniently charted using the Chart Wizard. Charting Excel data is discussed in Tutorial E.

Copying Summary Sheet Data to the Clipboard

You may want to put the summary sheet data into the Clipboard to use later in a Word document. To do that, use the following steps:

- Highlight the data range.
- Use Edit—Copy to put the graphic into the Clipboard.
- Assuming that you want to exit Excel, Select File—Save, File—Close, File—Exit Excel. (You may be asked whether you want to leave your data in the Clipboard—you do want to.)
- Open your Word document.
- Put your cursor where you want the upper-left part of the graphic to be positioned.
- Select Edit—Paste.

PRACTICE USING SCENARIO MANAGER

Suppose that you have an uncle who works for a large company. He has a good job and makes a decent salary ($80,000 a year, currently). He can retire from his company in 2011, when he will be 65. He would start drawing his pension then.

However, the company has an "early out" plan. Under this plan, the company asks employees to quit (called "pre-retirement"). The company then pays those employees a bonus in the year they quit and each year thereafter, up to the official retirement date, which is through the year 2010 for your uncle. Then, employees start to receive their actual pension—in your uncle's case, in 2011. This "early out" program would let your uncle leave the company before 2011. Until then, he could find a part-time hourly-wage job to make ends meet and then leave the workforce entirely in 2011.

The opportunity to leave early is open through 2010. Your uncle could stay with the company in 2005, then pre-retire any time in the years 2006 to 2010, getting the "early out" bonuses in those years. (Of course, if he retires in 2006, he would lose the 2005 bonus, and so on, all the way through 2010.

Another factor in your uncle's thinking is whether to continue belonging to his country club. He likes the club, but it is a real cash drain. The "early out" decision can be looked at each year, but the country club membership decision must be made now—if he does not withdraw in 2005, then he says he will stay a member (and incur costs) through 2010.

Your uncle has called you in to make a Scenario Manager spreadsheet model of his situation. Your spreadsheet would let him play "what if" with the pre-retirement and country club possibilities to see projected 2005-2010 personal finance results. He wants to know what "cash on hand" will be available for each year in the period with each scenario.

Complete the spreadsheet for your uncle. Your **SUMMARY OF KEY RESULTS**, **CALCULATIONS**, and **INCOME STATEMENT AND CASH FLOW STATEMENT** section cells must show values *by cell formula*. That is, in those areas, do not hard-code amounts. In any of your formulas, do not use the address of a cell if its contents are "NA." Set up your spreadsheet skeleton as shown in the figures that follow. Name your spreadsheet **UNCLE.xls**.

CONSTANTS Section

Your spreadsheet should have the constants shown in Figure C-21. An explanation of line items follows the figure.

	A	B	C	D	E	F	G	H
1	YOUR UNCLE'S EARLY RETIREMENT DECISION							
2	CONSTANTS	2004	2005	2006	2007	2008	2009	2010
3	CURRENT SALARY	80000	NA	NA	NA	NA	NA	NA
4	SALARY INCREASE FACTOR	NA	0.03	0.03	0.02	0.02	0.01	0.01
5	PART-TIME WAGES EXPECTED	NA	10000	10200	10500	10800	11400	12000
6	BUY OUT AMOUNT	NA	30000	25000	20000	15000	5000	0
7	COST OF LIVING (NOT RETIRED)	NA	41000	42000	43000	44000	45000	46000
8	COUNTRY CLUB DUES	NA	12000	13000	14000	15000	16000	17000

Figure C-21 CONSTANTS section values

- SALARY INCREASE FACTOR: Your uncle's salary at the end of 2004 will be $80,000. As you can see, raises are expected in each year—for example, a 3% raise is expected in 2005. If he does not retire, he would get his salary and the small raise in a year.
- PART-TIME WAGES EXPECTED: Your uncle has estimated his part-time wages if he were retired and working part time in the 2005-2010 period.
- BUY OUT AMOUNT: The company's pre-retirement "buy out" plan amounts are shown. If your uncle retires in 2005, he gets $30,000, $25,000, $20,000, $15,000, $5,000, and zero in the years 2005 to 2010, respectively. If he leaves in 2005, he gives up the $30,000 2005 payment, but would get $25,000, $20,000, $15,000, $5,000, and zero in the years 2006 to 2010, respectively.
- COST OF LIVING: Your uncle has estimated how much cash he needs to meet his living expenses, assuming that he continues to work for the company. His cost of living would be $41,000 in 2005, increasing each year thereafter.
- COUNTRY CLUB DUES: Country club dues are $12,000 for 2005. They increase each year thereafter.

INPUTS Section

Your spreadsheet should have the inputs shown in Figure C-22. An explanation of line items follows the figure.

	A	B	C	D	E	F	G	H
10	INPUTS	2004	2005	2006	2007	2008	2009	2010
11	RETIRED [R] or WORKING [W]	NA						
12	STAY IN CLUB? [Y] OR [N]	NA		NA	NA	NA	NA	NA

Figure C-22 INPUTS section

- RETIRED: Enter an "R" if your uncle retires in a year, or a "W" if he is still working. If he is working through 2010, the pattern **WWWWWW** should be entered. If his retirement is in 2005, the pattern **RRRRRR** should be entered. If he works for three years and then retires in 2008, the pattern **WWWRRR** should be entered.
- STAY IN CLUB: If your uncle stays in the club in 2005–2010, a "**Y**" should be entered. If the uncle is leaving the club in 2005, the letter "**N**" should be entered. The decision applies to all years.

SUMMARY OF KEY RESULTS Section

Your spreadsheet should show the results in Figure C-23.

	A	B	C	D	E	F	G	H
14	SUMMARY OF KEY RESULTS	2004	2005	2006	2007	2008	2009	2010
15	END-OF-THE-YEAR CASH ON HAND	NA						

Figure C-23 SUMMARY OF KEY RESULTS section

Each year's END-OF-THE-YEAR CASH ON HAND value is echoed from cells in the spreadsheet body.

CALCULATIONS Section

Your spreadsheet should calculate, by formula, the values shown in Figure C-24. Calculated amounts are used later in the spreadsheet. An explanation of line items follows the figure.

	A	B	C	D	E	F	G	H
17	CALCULATIONS	2004	2005	2006	2007	2008	2009	2010
18	TAX RATE	NA						
19	COST OF LIVING	NA						
20	YEARLY SALARY OR WAGES	80000						
21	COUNTRY CLUB DUES PAID	NA						

Figure C-24 CALCULATIONS section

- TAX RATE: Your uncle's tax rate depends on whether he is retired. Retired people have lower overall tax rates. If he retires in a year, your uncle's rate is expected to be 15% of income before taxes. In a year in which he works full time, the rate will be 30%.
- COST OF LIVING: In any year that your uncle continues to work for the company, his cost of living is what is shown in COST OF LIVING (NOT RETIRED) in the **CONSTANTS** section in Figure C-21. But if he chooses to retire, his cost of living is $15,000 less than the amount shown in the figure.
- YEARLY SALARY OR WAGES: If your uncle keeps working, his salary increases each year. The year-to-year percentage increases are shown in the **CONSTANTS** section. Thus, salary earned in 2005 would be more than that earned in 2004, salary earned in 2006 would be more than that earned in 2005, and so on. If your uncle retires in a certain year, he will make the part-time wages shown in the **CONSTANTS** section.
- COUNTRY CLUB DUES PAID: If your uncle leaves the club, the dues are zero each year; otherwise, the dues are as shown in the **CONSTANTS** section.

The INCOME STATEMENT AND CASH FLOW STATEMENT Section

This section begins with the cash on hand at the beginning of the year. This is followed by the income statement, concluding with the calculation of cash on hand at the end of the year. The format is shown in Figure C-25. An explanation of line items follows the figure.

	A	B	C	D	E	F	G	H
23	**INCOME STATEMENT AND CASH FLOW STATEMENT**	**2004**	**2005**	**2006**	**2007**	**2008**	**2009**	**2010**
24	BEGINNING-OF-THE-YEAR CASH ON HAND	**NA**						
25								
26	SALARY OR WAGES	**NA**						
27	BUY OUT INCOME	**NA**						
28	TOTAL CASH INFLOW	**NA**						
29	COUNTRY CLUB DUES PAID	**NA**						
30	COST OF LIVING	**NA**						
31	TOTAL COSTS	**NA**						
32	INCOME BEFORE TAXES	**NA**						
33	INCOME TAX EXPENSE	**NA**						
34	NET INCOME AFTER TAXES	**NA**						
35								
36	END-OF-THE-YEAR CASH ON HAND (BEGINNING-OF-THE-YEAR CASH PLUS NET INCOME AFTER TAXES)	30000						

Figure C-25 INCOME STATEMENT AND CASH FLOW STATEMENT section

- BEGINNING-OF-THE-YEAR CASH ON HAND: This is the END-OF-THE-YEAR CASH ON HAND at the end of the prior year.
- SALARY OR WAGES: This is a yearly calculation.
- BUY OUT INCOME: This is the year's "buy out" amount, if your uncle retires in the year.
- TOTAL CASH INFLOW: This is the sum of salary or part-time wages and "buy out" amounts.
- COUNTRY CLUB DUES PAID: This is a calculated amount.
- COST OF LIVING: This is a calculated amount.
- TOTAL COSTS: These outflows are the sum of the COST OF LIVING and COUNTRY CLUB DUES PAID.
- INCOME BEFORE TAXES: This amount is the TOTAL CASH INFLOW less TOTAL COSTS (outflows).
- INCOME TAX EXPENSE: This amount is zero if INCOME BEFORE TAXES is zero or less; otherwise, the calculated tax rate is applied to the INCOME BEFORE TAXES.
- NET INCOME AFTER TAXES: This is INCOME BEFORE TAXES, less TAX EXPENSE.
- END-OF-THE-YEAR CASH ON HAND: This is the BEGINNING-OF-THE-YEAR CASH plus the year's NET INCOME AFTER TAXES.

Scenario Manager Analysis

Set up the Scenario Manager and create a Scenario Summary sheet. Your uncle wants to look at the following four possibilities:

- Retire in 2005, staying in the club ("Loaf-In")
- Retire in 2005, leaving the club ("Loaf-Out")
- Work three more years, retire in 2008, staying in the club ("Delay-In")
- Work three more years, retire in 2008, leaving the club ("Delay-Out")

The output cell should be the 2010 (only) END-OF-THE-YEAR CASH ON HAND cell.

Your uncle will choose the option that yields the highest 2010 END-OF-THE-YEAR CASH ON HAND. You must look at your Scenario Summary sheet to see which strategy yields the highest amount.

To check your work, you should attain the values shown in Figure C-26. (You can use the labels Excel gives you in the left-most column or change the labels, as was done in Figure C-26.)

	A	B	C	D	E	F
1	Scenario Summary					
2			LOAF-IN	LOAF-OUT	DELAY-IN	DELAY-OUT
4	Changing Cells:					
5	RETIRE OR WORK, 2005	C11	R	R	W	W
6	RETIRE OR WORK, 2006	D11	R	R	W	W
7	RETIRE OR WORK, 2007	E11	R	R	W	W
8	RETIRE OR WORK, 2008	F11	R	R	R	R
9	RETIRE OR WORK, 2009	G11	R	R	R	R
10	RETIRE OR WORK, 2010	H11	R	R	R	R
11	IN CLUB 2005-2010?	C12	Y	N	Y	N
12	Result Cells (2010):					
13	END-OF-THE-YEAR CASH ON HAND	H15	-$68,400	$15,195	$8,389	$83,689

Figure C-26 Scenario Summary

Review of Excel Basics

In this section, you'll begin by reviewing how to perform some basic operations. Then, you'll work through some further cash flow calculations. Reading and working through this section will help you do the spreadsheet cases in this book.

Basic Operations

In this section, you'll review the following topics: formatting cells, showing Excel cell formulas, circular references, using the And and the Or functions in IF statements, and using nested IF statements.

Formatting Cells

You may have noticed that some data in this tutorial's first spreadsheet was centered in the cells. Here is how to perform that operation:

1. Highlight the cell range to format.
2. Select the Format menu option.
3. Select Cells—Alignment.
4. Choose Center for both Horizontal and Vertical.
5. Select OK.

It is also possible to put a border around cells. This treatment might be desirable for highlighting **INPUTS** section cells. To perform this operation:

1. Select Format—Cells—Border—Outline.
2. Choose the outline Style you want.
3. Select OK.

You can format numerical values for Currency format by selecting:

Format—Cells—Number—Currency.

You can format numerical values for decimal places using this procedure:

1. Select Format—Cells—Number tab—Number.
2. Select the desired number of decimal places.

Showing Excel Cell Formulas

If you want to see Excel cell formulas, follow this procedure:

1. Press the Ctrl key and the "back-quote" key (`) at the same time. (The back-quote orients from Seattle to Miami—on most keyboards, it is next to the exclamation-point key and shares the key with the tilde diacritic mark.)
2. To restore, press the Ctrl key and back-quote key again.

Understanding a Circular Reference

A formula has a circular reference if the *formula refers to the cell that the formula is already in*. Excel cannot evaluate such a formula, because the value of the cell is not yet known—but to do that evaluation, the value in the cell must be known! The reasoning is circular—hence the term "circular reference." Excel will point out circular references, if any exist, when you choose Open for a spreadsheet. Excel will also point out circular references as you insert them during development. Excel will be demonstrative about this by opening at least one Help window and by drawing arrows between cells involved in the offending formula. You can close the windows, but that will not fix the situation. You *must* fix the formula that has the circular reference if you want the spreadsheet to give you accurate results.

Here is an example. Suppose that the formula in cell C18 was =C18 – C17. Excel tries to evaluate the formula in order to put a value on the screen in cell C18. To do that, Excel must know the value in cell C18—but that is what it is trying to do in the first place. Can you see the circularity?

Using the "And" Function and the "Or" Function in =IF Statements

An =IF() statement has the following syntax:

=IF(test condition, result if test is True, result if test is False)

The test conditions in this tutorial's =IF statements tested only one cell's value. A test condition could test more than one cell's values.

Here is an example from this tutorial's first spreadsheet. In that example, selling price was a function of the economy. Assume, for the sake of illustration, that year 2004's selling price per unit depends on the economy *and* the purchase price outlook. If the economic outlook is optimistic *and* the company's purchase price outlook is down, then the selling price will be 1.10 times the prior year's price. Assume that in all other cases, the selling price will be 1.03 times the prior year's price. The first test requires two things to be true *at the same time*: C8 = "O" *AND* C9 = "D." So, the AND() function would be needed. The code in cell C17 would be as follows:

=IF(AND(C8 = "O", C9 = "D"), 1.10 * B17, 1.03 * B17)

On the other hand, the test might be this: If the economic outlook is optimistic *or* the purchase price outlook is down, then the selling price will be 1.10 times the prior year's price. Assume that in all other cases, the selling price will be 1.03 times the prior year's price. The first test requires *only one of* two things to be true: C8 = "O" *or* C9 = "D". Thus, the OR() function would be needed. The code in cell C17 would be:

=IF(OR(C8 = "O", C9 = "D"), 1.10 * B17, 1.03 * B17)

Using IF() Statements Inside IF() Statements

An =IF() statement has this syntax:

=IF(test condition, result if test is True, result if test is False)

In the examples shown thus far, only two courses of action were possible, so only one test has been needed in the =IF() statement. There can be more courses of action than two, however, and this requires that the "result if test is False" clause needs to show further testing. Let's look at an example.

Assume again that the 2005 selling price per unit depends on the economy and the purchase price outlook. Here is the logic: (1) If the economic outlook is optimistic *and* the purchase price outlook is down, then the selling price will be 1.10 times the prior year's price. (2) If the economic outlook is optimistic *and* the purchase price outlook is up, then the selling price will be 1.07 times the prior year's price. (3) In all other cases, the selling price will be 1.03 times the prior year's price. The code in cell C17 would be:

=IF(AND(C8 = "O", C9 = "D"), 1.10 * B17,
IF(AND(C8 = "O", C9 = "U"), 1.07 * B17, 1.03 * B17))

The first =IF() tests to see if the economic outlook is optimistic and the purchase price outlook is down. If not, further testing is needed to see whether the economic outlook is optimistic and the purchase price outlook is up, or whether some other situation prevails.

NOTE

Be sure to note the following:

- The line is broken in the previous example because the page is not wide enough, but in Excel, the formula would appear on one line.
- The embedded "IF" is not preceded by an equals sign.

Example: Borrowing and Repayment of Debt

The Scenario Manager cases require you to account for money that the company borrows or repays. Borrowing and repayment calculations are discussed next. At times you are asked to think about a question and fill in the answers. Correct responses are at the end of this section.

To do the Scenario Manager cases, you must assume two things about a company's borrowing and repayment of debt. First, assume that the company wants to have a certain minimum cash level at the end of a year (and thus to start the next year). Assume that a bank will provide a loan to reach the minimum cash level if year-end cash falls short of that level.

Here are some numerical examples to test your understanding. Assume that NCP stands for "net cash position" and equals beginning-of-the-year cash plus net income after taxes for the year. The NCP is the cash available at year end, before any borrowing or repayment. Compute the amounts to borrow in the three examples in Figure C-27.

Example	NCP	Minimum Cash Required	Amount to Borrow
1	50,000	10,000	?
2	8,000	10,000	?
3	–20,000	10,000	?

Figure C-27 Examples of borrowing

Assume that a company would use its excess cash at year end to pay off as much debt as possible, without going below the minimum-cash threshold. "Excess cash" would be the NCP *less* the minimum cash required on hand—amounts over the minimum are available to repay any debt.

In the examples shown in Figure C-28, compute excess cash and then compute the amount to repay. You may also want to compute ending cash after repayments as well, to aid your understanding.

Example	NCP	Minimum Cash Required	Beginning-of-the-Year Debt	Repay	Ending Cash
1	12,000	10,000	4,000	?	?
2	12,000	10,000	10,000	?	?
3	20,000	10,000	10,000	?	?
4	20,000	10,000	0	?	?
5	60,000	10,000	40,000	?	?
6	–20,000	10,000	10,000	?	?

Figure C-28 Examples of repayment

In this section's Scenario Manager cases, your spreadsheet will need two bank financing sections beneath the **INCOME STATEMENT AND CASH FLOW STATEMENT** section:

1. The first section will calculate any needed borrowing or repayment at the year's end to compute year-end cash.
2. The second section will calculate the amount of debt owed at the end of the year, after borrowing or repayment of debt.

The first new section, in effect, extends the end-of-year cash calculation, which was shown in Figure C-13. Previously, the amount equaled cash at the beginning of the year plus the year's net income. Now, the calculation will include cash obtained by borrowing and cash repaid. Figure C-29 shows the structure of the calculation.

	A	B	C	D
30	NET CASH POSITION (NCP) BEFORE BORROWING AND REPAYMENT OF DEBT (BEGINNING-OF-THE-YEAR CASH PLUS NET INCOME AFTER TAXES)	NA		
31	PLUS: BORROWING FROM BANK	NA		
32	LESS: REPAYMENT TO BANK	NA		
33	EQUALS: END-OF-THE-YEAR CASH ON HAND	10000		

Figure C-29 Calculation of END-OF-THE-YEAR CASH ON HAND

The heading in cell A30 was previously END-OF-THE-YEAR CASH ON HAND. But BORROWING increases cash and REPAYMENT OF DEBT decreases cash. So, END-OF-THE-YEAR CASH ON HAND is now computed two rows down (in C33 for year 2005, in the example). The value in row 30 must be a subtotal for the BEGINNING-OF-THE-YEAR CASH ON HAND plus the year's NET INCOME AFTER TAXES. We call that subtotal the NET CASH POSITION (NCP) BEFORE BORROWING AND REPAYMENT OF DEBT.

(*Note*: Previously, the formula in cell C22 for BEGINNING-OF-THE-YEAR CASH ON HAND was =B30. Now, that formula would be =B33. It would be copied to the right, as before, for the next year.)

That second new section would look like what is shown in Figure C-30.

	A	B	C	D
35	**DEBT OWED**	**2004**	**2005**	**2006**
36	BEGINNING-OF-THE-YEAR DEBT OWED	NA		
37	PLUS: BORROWING FROM BANK	NA		
38	LESS: REPAYMENT TO BANK	NA		
39	EQUALS: END-OF-THE-YEAR DEBT OWED	15000		

Figure C-30 DEBT OWED section

The second new section computes end-of-year debt and is called DEBT OWED. At the end of 2004, $15,000 was owed. END-OF-THE-YEAR DEBT OWED equals the BEGINNING-OF-THE-YEAR DEBT OWED, plus any new BORROWING FROM BANK (which increases debt owed), less any REPAYMENT TO BANK (which reduces it). So, in the example, the formula in cell C39 would be:

=C36 + C37 – C38

Assume that the amounts for BORROWING FROM BANK and REPAYMENT TO BANK are calculated in the first new section. Thus, the formula in cell C37 would be: =C31. The formula in cell C38 would be: =C32. (BEGINNING-OF-THE-YEAR DEBT OWED is equal to the debt owed at the end of the prior year, of course. The formula in cell C36 for BEGINNING-OF-THE-YEAR DEBT OWED would be an echoed formula. *Can you see what it would be*? It's an exercise for you to complete. *Hint*: The debt owed at the beginning of a year equals the debt owed at the end of the prior year.)

Now that you have seen how the borrowing and repayment data is shown, the logic of the borrowing and repayment formulas can be discussed.

Calculation of BORROWING FROM BANK

The logic of this in English is:

If (cash on hand before financing transactions is greater than the
minimum cash required, then borrowing is not needed;
otherwise, borrow enough to get to the minimum).

Or (a little more precisely):

If (NCP is greater than the minimum cash required,
then BORROWING FROM BANK = 0; otherwise,
borrow enough to get to the minimum).

Suppose that the desired minimum cash at year end is $10,000, and that value is a constant in your spreadsheet's cell C6. Assume the NCP is shown in your spreadsheet's cell C30. Our formula (getting closer to Excel) would be as follows:

IF(NCP > Minimum Cash, 0; otherwise, borrow enough to get to the minimum).

Tutorial C

You have cell addresses that stand for NCP (cell C30) and Minimum Cash (C6). To develop the formula for cell C31, substitute these cell addresses for NCP and Minimum Cash. The harder logic is that for the "otherwise" clause. At this point, you should look ahead to the borrowing answers at the end of this section, Figure C-31. In Example 2, $2,000 had to be borrowed. Which cell was subtracted from which other cell to calculate that amount? Substitute cell addresses in the Excel formula for the year 2004 borrowing formula in cell C31:

=IF(>=   , 0 ,   –   )

The Answer is at the end of this section, Figure C-33.

Calculation of REPAYMENT TO BANK

The logic of this in English is:

IF(beginning of year debt = 0, repay 0 because nothing is owed, but
    IF(NCP is less than the min, repay zero, because you must *borrow*, but
        IF(extra cash equals or exceeds the debt, repay the whole debt,
            ELSE (to stay above the min, repay only the extra cash))))

Look at the following formula. Assume the repayment will be in cell C32. Assume debt owed at the beginning of the year is in cell C36, and that minimum cash is in cell C6. Substitute cell addresses for concepts to complete the formula for year 2005 repayment. (Clauses are on different lines because of page width limitations.)

=IF(= 0, 0,
    IF(<=   , 0,
        IF((–   ) >=   ,
            (–   )))).

The answer is at the end of this section, in Figure C-34.

Answers to Questions About Borrowing and Repayment Calculations

Figures C-31 and C-32 give the answers to the questions posed about borrowing and repayment calculations.

Example	NCP	Minimum Cash Required	Borrow	Comments
1	50,000	10,000	Zero	NCP > Min.
2	8,000	10,000	2,000	Need 2000 to get to Min. 10,000 – 8,000.
3	–20,000	10,000	30,000	Need 30000 to get to Min. 10,000 – (–20,000).

Figure C-31 Answers to examples of borrowing

Example	NCP	Minimum Cash Required	Beginning-of-the-Year Debt	Repay	Ending Cash
1	12,000	10,000	4,000	2,000	10,000
2	12,000	10,000	10,000	2,000	10,000
3	20,000	10,000	10,000	10,000	10,000
4	20,000	10,000	0	0	20,000
5	60,000	10,000	40,000	40,000	20,000
6	–20,000	10,000	10,000	NA	NA

Figure C-32 Answers to examples of repayment

Some notes about the repayment calculations shown in Figure C-32 follow.

- In Examples 1 and 2, only $2,000 is available for debt repayment (12,000 – 10,000) to avoid going below the minimum cash.
- In Example 3, cash available for repayment is $10,000 (20,000 – 10,000), so all beginning debt can be repaid, leaving the minimum cash.
- In Example 4, there is no debt owed, so no debt need be repaid.
- In Example 5, cash available for repayment is $50,000 (60,000 – 10,000), so all beginning debt can be repaid, leaving more than the minimum cash.
- In Example 6, no cash is available for repayment. The company must borrow.

Figures C-33 and C-34 show the calculations for borrowing and repayment of debt.

```
=IF( C30 >= C6, 0, C6 – C30)
```

Figure C-33 Calculation of borrowing

```
=IF( C36 = 0, 0, IF( C30 <= C6, 0, IF( (C30 – C6) >= C36, C36, ( C30 – C6) )))
```

Figure C-34 Calculation of repayment

Saving Files After Using Microsoft Excel

As you work, save periodically (File—Save). If you want to save to a diskette, choose Drive A:. At the end of a session, save your work using this three-step procedure:

1. Save the file, using File—Save. If you want to save to a diskette, choose Drive A:.

2. Use File—Close to tell Windows to close the file. If saving to a diskette, make sure it is still in Drive A: when you close. If you violate this rule, you may lose your work!
3. Exit from Excel to Windows by selecting File—Exit. In theory, you may exit from Excel back to Windows after you have saved a file (short-cutting the File—Close step), but that is not a recommended shortcut.

6

CASE

The POURQUOI Company Advertising Campaign Decision

DECISION SUPPORT USING EXCEL

PREVIEW

The POURQUOI Company has two divisions, Division A and Division B. Each division makes and sells its own product. POURQUOI wants to run a single advertising campaign for the company as a whole, promoting both Division A's and B's products. Two different advertising campaigns are under consideration, but they affect division sales differently. In this case, you will use Excel to help management decide which campaign would be best for the whole company in the coming years.

PREPARATION

- Review spreadsheet concepts discussed in class and/or in your textbook.
- Complete any exercises that your instructor assigns.
- Complete any part of Tutorial C that your instructor assigns, or refer to it as necessary.
- Review file-saving procedures for Windows programs. These are discussed in Tutorial C.
- Refer to Tutorial E as necessary.

BACKGROUND

The POURQUOI Company has two divisions, Division A and Division B. Each makes and sells its own distinct product. Effective advertising is needed to support the competitiveness of each product in its marketing niche.

Advertising is done at the corporate level, however, and managers want to run a single POURQUOI campaign to promote both divisions' products.

A company's advertising program must be updated every so often, and POURQUOI likes to use the same advertising theme in a four-year campaign cycle. Assume that it is the end of 2004, and POURQUOI is now working with its ad agency, Purple Penguin Communications, on the next four-year campaign for 2005–2008. Advertising will be featured in print and on radio and television. POURQUOI managers are considering two campaigns: a "Warm and Fuzzy" campaign and a "Hard Edged" campaign.

In the Warm and Fuzzy campaign, POURQUOI products would be extolled in scenes that feature puppy dogs, the American flag, fields of grain, apple pie, and other soothing backdrops. In the Hard Edged campaign, POURQUOI products would be extolled by a loud, caustic comedian and his crazy girlfriend. Focus group results show that each campaign's ads elicit a strong emotional response in audiences.

The emotional responses are not the same, of course, and each seems destined to stimulate both divisions' sales. However, the Warm and Fuzzy campaign seems likely to benefit Division A's sales more than B's, while the Hard Edged campaign seems likely to have the opposite effect. The impact of each campaign on unit sales, product prices, and product costs in different economic scenarios has been estimated.

POURQUOI management has not yet decided which campaign to use. Management wants to see a forecasted income statement and cash flow statement for 2005 to 2008. The forecast must be based on what happened in 2004, on how good the economy may be in the period, and on the effects of the two ad campaigns. You must make a DSS in Excel that lets POURQUOI's management answer this question: Given an estimate of the economy and given the two ad campaigns, what will POURQUOI financial results be in 2005–2008?

In your DSS, the inputs are for (1) two states of the economy in the four-year period: Good and Neutral, and (2) two possible ad campaigns: Warm and Fuzzy or Hard Edged. Your DSS lets the POURQUOI Company management play "what if" with the inputs, see the results, then decide which ad campaign to use.

The impacts of the inputs on key operating variables are complex and are discussed more completely next.

Units Sold and Selling Prices

First, let's look at the number of units sold per year by division, and then at selling prices of units by division.

Division A: Number of Units Sold per Year

Sales are a function of the expected economy and of the ad campaign. Figure 6-1 shows the percentage increase in units sold in a year, in four possible situations.

Percentage Increase in Division A Units Sold		
	Good Economy	**Neutral Economy**
Warm and Fuzzy	10%	2%
Hard Edged	4%	1%

Figure 6-1 Percentage increase in Division A's units sold

Here is an explanatory example: If a "Good" economy is expected and the Warm and Fuzzy campaign is used, Division A would expect to sell 10% more units in a year than in the prior year. Thus, assuming a "Good" 2005 economy and the Warm and Fuzzy campaign, Division A's 2005 units sold would be 110% of 2004's units sold.

Division B: Number of Units Sold per Year

Sales for Division B are also a function of the expected economy and of the ad campaign. Figure 6-2 shows the percentage increase in units sold in a year, in four possible situations.

Percentage Increase in Division B Units Sold		
	Good Economy	**Neutral Economy**
Warm and Fuzzy	3%	7%
Hard Edged	5%	12%

Figure 6-2 Percentage increase in Division B's units sold

Division A: Unit Selling Price

Unit selling price is a function of the expected economy and of the ad campaign. Figure 6-3 shows the percentage increase in unit selling price in a year for Division A, in four possible situations.

Percentage Increase in Division A Unit Selling Price		
	Good Economy	**Neutral Economy**
Warm and Fuzzy	6%	2%
Hard Edged	4%	1%

Figure 6-3 Percentage increase in Division A's unit selling price

Explanatory example: If a "Good" economy is expected and the Warm and Fuzzy campaign used, Division A would expect its selling price to increase 6% over the prior year's selling price. Thus, assuming a "Good" 2004 economy and the Warm and Fuzzy campaign, Division A's 2005 selling price would be 106% of 2004's selling price.

Division B: Unit Selling Price

Division B's unit selling price is also a function of the expected economy and of the ad campaign. Figure 6-4 shows the percentage increase in Division B's unit selling price in a year, in four possible situations.

Percentage Increase in Division B Unit Selling Price

	Good Economy	Neutral Economy
Warm and Fuzzy	8%	2%
Hard Edged	4%	4%

Figure 6-4 Percentage increase in Division B's unit selling price

Cost to Make and Sell a Unit

In this section, let's look at each division's cost to make and sell a unit of product.

Division A: Cost to Make and Sell a Unit

The base cost to make and sell a unit is influenced only by the expected economy. No other factors come into play. Figure 6-5 shows the percentage increase in unit cost in a year in the two economies.

Percentage Increase in Division A Unit Cost

Good Economy	Neutral Economy
5%	2%

Figure 6-5 Percentage increase in Division A's unit cost

Explanatory example: If a "Good" economy is expected, Division A would expect its unit cost to increase 5% each year over the prior year. Thus, assuming a "Good" 2005 economy, Division A's 2005 unit cost would be 105% of 2004's unit cost, and so on.

Division B: Cost to Make and Sell a Unit

Division B's cost to make and sell a unit is also only a function of the expected economy. Figure 6-6 shows the percentage increase in unit cost in each year in the two economies.

Percentage Increase in Division B Unit Cost

Good Economy	Neutral Economy
3%	1%

Figure 6-6 Percentage increase in Division B's unit cost

Interest Rate on Debt Owed

Interest rate is influenced only by the expected economy. Given the two economies, Figure 6-7 shows the expected interest rate in a year.

Interest Rate in a Year	
Good Economy	**Neutral Economy**
8%	6%

Figure 6-7 Interest rate in year

Measuring Advertising Effectiveness

Now, let's look at the company's method of measuring the effectiveness of its advertising.

Members of POURQUOI's marketing management are convinced that effectiveness should be measured by the "return on sales" over the life of the campaign. Return on sales is a percentage computed by dividing net income after taxes by sales (revenue). The following amounts must be computed each year in the four-year period:

- Net income to date: This is the accumulated POURQUOI net income after taxes. In 2005, it is 2005's net income. In 2006, it is 2005's net income plus 2006's net income, and so forth, through 2008.
- POURQUOI sales to date: This is the accumulated POURQUOI total sales (revenue). Amounts are accumulated in the same way as for net income to date, through 2008.
- Return on sales to date: For each year, net income to date is divided by sales to date to give the return on sales to date. In 2008, this will be the return on sales for the entire campaign.

POURQUOI management thinks that the ad campaign that gives the best four-year return on sales, given the economic scenario, is the campaign that it should run.

Assignment 1 Creating a Spreadsheet for Decision Support

In this assignment, you will produce a spreadsheet that models a business decision. In Assignment 2, you will use the spreadsheet to gather data and then write a memorandum to POURQUOI Marketing management that explains your recommended action. In Assignment 3, you will prepare an oral presentation of your recommendations.

Now, you will begin creating a spreadsheet model of the ad campaign decision. The model is an income statement and cash flow forecast for the years 2005–2008. Your spreadsheet should have the sections that follow. You will be shown how each section should be set up before entering cell formulas.

- **CONSTANTS**
- **INPUTS**
- **SUMMARY OF KEY RESULTS**
- **CALCULATIONS**
- **INCOME STATEMENT AND CASH FLOW STATEMENT**
- **DEBT OWED**

A discussion of each spreadsheet section follows. The discussion details (1) how data in each section should be organized, and (2) the logic of the formula in the section's cells. *When you type in the spreadsheet skeleton, follow the order given here.*

CONSTANTS Section

Your spreadsheet should have the constants shown in Figure 6-8. An explanation of the line items follows the figure.

	A	B	C	D	E	F
1	POURQUOI COMPANY ADVERTISING CAMPAIGN DECISION					
2	CONSTANTS	2004	2005	2006	2007	2008
3	TAX RATE	NA	0.30	0.31	0.32	0.33
4	MINIMUM CASH NEEDED TO START YEAR	NA	10000	10000	10000	10000
5	WARM AND FUZZY CAMPAIGN COST	NA	80000	84000	90000	100000
6	HARD EDGED CAMPAIGN COST	NA	50000	60000	70000	80000

Figure 6-8 CONSTANTS section

- The TAX RATE expected in the years is expected to increase slightly, as you can see.
- POURQUOI needs $10,000 to start each year. This is called the MINIMUM CASH NEEDED TO START YEAR.
- The cost of the two advertising campaigns is shown. The costs go up each year.

INPUTS Section

Your spreadsheet should have the inputs shown in Figure 6-9. An explanation of the line items follows the figure.

	A	B	C	D	E	F
8	INPUTS	2004	2005	2006	2007	2008
9	ECONOMIC OUTLOOK (G = GOOD, N = NEUTRAL)	NA				
10	AD CAMPAIGN (F = WARM AND FUZZY, H = HARD EDGED)	NA		NA	NA	NA

Figure 6-9 INPUTS section

- If the economic outlook in the year is expected to be "Good," the user enters a **G**. If the economic outlook is expected to be "Neutral," the user enters an **N**. The same value could be entered for each year. Different patterns could be entered as well. An "improving economy" pattern might be **NNGG**.
- If the company adopts the Warm and Fuzzy campaign, an **F** should be entered. If the company adopts the Hard Edged campaign, an **H** should be entered. (*Note*: The value applies to all four years.)

Your instructor may tell you to apply Conditional Formatting to the input cells, so that out-of-bounds values are highlighted in some way (for example, in red type and/or in boldface type). If so, use the Excel Help system to determine how to use Conditional Formatting for this purpose, or follow the instructions provided by your instructor. (*Hint*: For the advertising campaign input, you will need to specify a formula, and the formula must use Excel's =AND() function.)

Case 6

SUMMARY OF KEY RESULTS Section

Your spreadsheet should resemble Figure 6-10.

	A	B	C	D	E	F
12	SUMMARY OF KEY RESULTS	2004	2005	2006	2007	2008
13	NET INCOME AFTER TAXES	NA				
14	END-OF-THE-YEAR CASH ON HAND	NA				
15	END-OF-THE-YEAR DEBT OWED	NA				
16	RETURN ON SALES TO DATE	NA				

Figure 6-10 SUMMARY OF KEY RESULTS section

For each year, your spreadsheet should show (1) net income after taxes for the year, (2) cash on hand at the end of the year, (3) debt owed to the company's banker at the end of the year, and (4) the return on sales to date. These values are all computed elsewhere in the spreadsheet and echoed to this section.

To enhance readability, you should format cells in this section for zero decimal places, except for the return on sales, which should be formatted for three decimal places.

CALCULATIONS Section

You should calculate the intermediate results, which are then used in the **INCOME STATEMENT AND CASH FLOW STATEMENT** section. Calculations, shown in Figure 6-11, are based on input values and/or on year 2004 values. When called for, use absolute addressing. An explanation of the line items follows the figure.

For readability in this section, you should format selling prices and cost of goods sold for two decimal places, and all other cells for zero decimal places.

	A	B	C	D	E	F
18	CALCULATIONS	2004	2005	2006	2007	2008
19	DIVISION A UNITS SOLD	2000				
20	DIVISION B UNITS SOLD	3000				
21	DIVISION A UNIT SELLING PRICE	100				
22	DIVISION B UNIT SELLING PRICE	117				
23	DIVISION A UNIT COST OF GOODS SOLD	87				
24	DIVISION B UNIT COST OF GOODS SOLD	112				
25	INTEREST RATE	NA				
26	NET INCOME AFTER TAXES TO DATE	NA				
27	REVENUE (SALES) TO DATE	NA				
28	RETURN ON SALES TO DATE	NA				

Figure 6-11 CALCULATIONS section

- UNITS SOLD and UNIT SELLING PRICE are both a function of the economy and the ad campaign. In each case, four combinations of values are possible.
- UNIT COST OF GOODS SOLD values are a function of the expected economy.
- The INTEREST RATE prevailing in a year is a function of the expected economy.

- NET INCOME AFTER TAXES TO DATE is the accumulated POURQUOI net income thus far. In 2005, net income is only 2005's net income. In 2006, net income is 2005's net income plus 2006's net income, and so forth, until 2008.
- Similarly, REVENUE (SALES) TO DATE is the accumulated POURQUOI revenue so far. In 2005, revenue is only 2005's revenue. In 2006, it is 2005's revenue plus 2006's revenue, and so forth, until 2008. Note that revenue is summed for the two divisions.
- To get the RETURN ON SALES TO DATE value, net income to date is divided by sales to date.

INCOME STATEMENT AND CASH FLOW STATEMENT Section

The forecast for net income and cash flow starts with the cash on hand at the beginning of the year. This is followed by the income statement and concludes with the calculation of cash on hand at year-end. For readability in this section, results should be formatted for zero decimals (i.e., no pennies). Your spreadsheet section should look like the one shown in Figures 6-12 and 6-13. A discussion of line items follows the figures.

	A	B	C	D	E	F
30	**INCOME STATEMENT AND CASH FLOW STATEMENT**	**2004**	**2005**	**2006**	**2007**	**2008**
31	BEGINNING-OF-THE-YEAR CASH ON HAND	**NA**				
32						
33	REVENUE (SALES)	-				
34	DIVISION A REVENUE (SALES)	**NA**				
35	DIVISION B REVENUE (SALES)	**NA**				
36	TOTAL REVENUE	**NA**				
37	COSTS AND EXPENSES	-				
38	DIVISION A COST OF GOODS SOLD	**NA**				
39	DIVISION B COST OF GOODS SOLD	**NA**				
40	ADVERTISING CAMPAIGN COST	**NA**				
41	TOTAL COSTS AND EXPENSES	**NA**				
42	INCOME BEFORE INTEREST AND TAXES	**NA**				
43	INTEREST EXPENSE	**NA**				
44	INCOME BEFORE TAXES	**NA**				
45	INCOME TAX EXPENSE	**NA**				
46	NET INCOME AFTER TAXES	**NA**				

Figure 6-12 INCOME STATEMENT AND CASH FLOW STATEMENT section

- BEGINNING-OF-THE-YEAR CASH ON HAND is the cash on hand at the end of the *prior* year.
- DIVISION REVENUE (SALES) is a function of the division's units sold and unit selling price.
- TOTAL REVENUE is the sum of the two divisions' revenues.
- DIVISION COST OF GOODS SOLD is a function of the units sold and the unit cost of goods sold.
- ADVERTISING CAMPAIGN COST depends on the campaign chosen, which is an input value.
- TOTAL COSTS AND EXPENSES is the sum of divisional cost of goods sold and the cost of the ad campaign.

- INCOME BEFORE INTEREST AND TAXES is a subtotal for the difference between total revenue and total costs and expenses.
- INTEREST EXPENSE is based on the year's interest rate (a calculation) and beginning-of-the-year debt owed.
- INCOME BEFORE TAXES is the difference between income before interest and taxes and the interest expense.
- INCOME TAX EXPENSE is zero if income before taxes is zero or less; otherwise, apply the tax rate to income before taxes to determine the tax expense.
- NET INCOME AFTER TAXES is income before taxes less income tax expense.

Continuing this statement, line items for the year-end cash calculation are discussed. In Figure 6-13, column B is for year 2004, column C for 2005, and so on.

	A	B	C	D	E	F
48	NET CASH POSITION (NCP) BEFORE BORROWING AND REPAYMENT OF DEBT (BEG YR CASH + NET INCOME)	**NA**				
49	ADD: BORROWING FROM BANK	**NA**				
50	LESS: REPAYMENT TO BANK	**NA**				
51	EQUALS: END-OF-THE-YEAR CASH ON HAND	10000				

Figure 6-13 END-OF-THE-YEAR CASH ON HAND calculation

- Year 2004 values are mostly NA, except $10,000 END-OF-THE-YEAR CASH ON HAND. This amount happens to equal POURQUOI's minimum cash to start a year.
- THE NET CASH POSITION (NCP) at the end of a year equals the cash at the beginning of a year plus the year's net income.
- POURQUOI's bank will lend them enough money at year-end to get to the minimum cash. If the NCP is less than minimum cash, then POURQUOI must borrow enough cash to reach the minimum.
- If the NCP is more than the minimum cash at the end of a year and there is outstanding debt, then as much debt as possible should be repaid (but not to take POURQUOI below the minimum cash required).
- END-OF-THE-YEAR CASH ON HAND equals the NCP plus any borrowing, less any repayment.

DEBT OWED Section

Your spreadsheet body ends with a calculation of debt owed at year-end, as shown in Figure 6-14. An explanation of line items follows the figure.

	A	B	C	D	E	F
53	**DEBT OWED**	**2004**	**2005**	**2006**	**2007**	**2008**
54	BEGINNING-OF-THE-YEAR DEBT OWED	**NA**				
55	ADD: BORROWING FROM BANK	**NA**				
56	LESS: REPAYMENT TO BANK	**NA**				
57	EQUALS: END-OF-THE-YEAR DEBT OWED	50000				

Figure 6-14 DEBT OWED section

- Year 2004 values are mostly NA, except that $50,000 was owed to the bank at year-end.
- Cash owed to the bank at the beginning of a year (BEGINNING-OF-THE-YEAR DEBT OWED) equals cash owed to the bank at the end of the prior year.
- Amounts borrowed and repaid have been calculated and can be echoed to this section.
- The amount owed at the end of a year equals the amount owed at the beginning of a year, plus any borrowing and less any repayment.

ASSIGNMENT 2 USING THE SPREADSHEET FOR DECISION SUPPORT

You will now complete the case by (1) using the spreadsheet to gather the data needed to make the ad campaign decision, and (2) documenting your recommendation in a memorandum.

Assignment 2A: Using the Spreadsheet to Gather Data

You have built the spreadsheet to model the ad campaign decision. Assume that POURQUOI Company's management (and the company's banker!) are interested in these six scenarios:

- The economy is "Good" in all four years (**GGGG** input pattern), and the Warm and Fuzzy ad campaign is used (**F** input). Call the scenario *Good-Fuzzy*.
- The economy is "Good" in all four years, and the Hard Edged ad campaign is used. Call the scenario *Good-Hard*.
- The economy is "Neutral" in all four years (**NNNN** pattern), and the Warm and Fuzzy ad campaign is used. Call the scenario *Neutral-Fuzzy*.
- The economy is "Neutral" in all four years, and the Hard Edged ad campaign is used. Call the scenario *Neutral-Hard*.
- The economy is "improving" in all four years (**NNGG** pattern), and the Warm and Fuzzy ad campaign is used. Call the scenario *Improving-Fuzzy*.
- The economy is "improving" in all four years (**NNGG** pattern), and the Hard Edged ad campaign is used. Call the scenario *Improving-Hard*.

Now run "what if" scenarios with the possible six sets of input values. The way you do this depends on whether your instructor has told you to chart 2008 results and/or to use the Scenario Manager.

- If your instructor has told you *not* to use the Scenario Manager, you must now manually enter the input value combinations. Note the results for each on a piece of paper. (The results will show in the **SUMMARY OF KEY RESULTS** section as you work). Additionally, if your instructor has told you to chart the 2008 results, manually enter these into a separate spreadsheet section beneath the DEBT OWED section. (You will need room for each of the six 2008 alternatives.) The chart would show the scenario designator (e.g., Good-Fuzzy) and the related return on sales through 2008.
- If your instructor has told you to use the Scenario Manager, perform the procedures set forth in Tutorial C to set up and run the Scenario Manager. Record the six possible scenarios. The changing cells are the cells used to input the economy and the ad campaign values. (In the Scenario Manager, you can enter non-contiguous input ranges as follows: C9..F9, C10.) The output cell is for 2008 return on sales to date only. If your instructor told you to chart the 2008 results, you can chart the Summary Sheet values, which are nicely arrayed for that purpose.

- In either case, when you are done gathering data, print the entire workbook (including any charts on their own sheets). Then save the **.xls** file. **AD.xls** is recommended as a file name.

Assignment 2B: Documenting Your Recommendations in a Memorandum

Open MS Word and write a brief memorandum to POURQUOI's Director of Marketing, who wants to know which ad campaign strategy seems best. Assume that the director would consider an ad campaign only if its 2008 return on sales to date was positive. The ad campaign may get off to a slow start, but *by 2008, company management wants to see a return on sales to date of at least 3%*. Thus, the director needs to know the answer to these questions:

1. Given a "Good" economy, which campaign(s) produce a positive return on sales to date in 2008?
2. Given a "Neutral" economy, which campaign(s) produce a positive return on sales to date in 2008?
3. Given an "improving" economy, which campaign(s) produce a positive return on sales to date in 2008?
4. Is it possible not to earn a positive return on sales to date in 2008 in *any* of the six scenarios?
5. Which campaign (if any) produces a return on sales to date in 2008 that is greater than 3%?

If the director knows the answer to her questions, she can decide which campaign to use, after POURQUOI's economist tells her which kind of economy to expect in the four-year period. When creating your memo, observe the following guidelines:

- Your memorandum should have a proper heading (DATE / TO / FROM / SUBJECT). You may wish to use a Word memo template (File—New, click Memos, choose Contemporary Memo, and click OK).
- In the first paragraph, tell the Director which ad campaign strategy is best in each kind of economy. If more than one strategy seems "tied" for best, state that and explain your reasoning.
- In view of management's need for a 3% return on sales, tell the Director which campaign you recommend for the company. If no campaign meets the 3% goal, tell management that fact, and identify the best of a poor set of alternatives.
- Support your statement graphically as your instructor requires: (1) If you used the Scenario Manager, go back to Excel and put a copy of the Summary Sheet results into the Windows Clipboard. Then, in Word, copy this from the Clipboard. (Tutorial C refers to this procedure.) (2) If you made a chart in Excel, copy it from Excel to Word in the same way. (3) Make a summary table in Word and position it after the memo's first paragraph. This procedure is described next.

Enter a table into Word, using the following procedure:

1. Select the **Table** menu option, click **Insert**, and then click **Table**.
2. Enter the number of rows and columns.
3. Select **AutoFormat** and choose **Table Grid 1**.
4. Select **OK**, and then select **OK** again.

Your table should resemble the format of the table shown in Figure 6-15.

2008 Return on Sales to Date			
Advertising campaign	*Good economy in all four years*	*Neutral economy in all four years*	*Improving economy for the four years*
Warm and Fuzzy			
Hard Edged			

Figure 6-15 Format of table to insert in memorandum

ASSIGNMENT 3 GIVING AN ORAL PRESENTATION

Your instructor may request that you also present your analysis and recommendations in an oral report. If so, assume that POURQUOI has accepted the analysis and recommendations that you set forth in your memo. The marketing director has asked you to give a presentation explaining your recommendation to a joint meeting of management and the company's bankers who hold promissory notes for debt owed. Prepare to explain your analysis and recommendation to the group in 10 minutes or less. Use visual aids, such as PowerPoint®, or handouts that you think are appropriate. Tutorial E in this casebook has guidance on how to prepare an oral presentation.

DELIVERABLES

Assemble the following deliverables for your instructor:

1. Memorandum
2. Spreadsheet printouts
3. Diskette, which should have your Word memo file and your Excel spreadsheet file

Staple the printouts together, with the memo on top. If there is more than one **.xls** file on your diskette, write your instructor a note, stating the name of your model's **.xls** file.

7 CASE

The Baseball Team Purchase Decision

Decision Support Using Excel

Preview

Your hometown major league baseball team is for sale. A prominent and wealthy person would be capable of buying the team; however, she would have to borrow the purchase price. In this case, you will use Excel to see whether the team will be profitable enough to allow the prospective owner to pay off the bank debt in a reasonable amount of time.

Preparation

- Review spreadsheet concepts discussed in class and/or in your textbook.
- Complete any exercises that your instructor assigns.
- Complete any part of Tutorial C that your instructor assigns, or refer to it as necessary.
- Review file-saving procedures for Windows programs. These are discussed in Tutorial C.
- Refer to Tutorial E as necessary.

BACKGROUND

Your hometown baseball team has been generally successful, although it has been many, many years since the team has won the World Series. The team has always been profitable because its games are well attended. In addition, the team owns its stadium—usually, major league baseball stadiums are owned by a city. Owning the stadium means the team pays no rent and does not share concession revenue with the city (parking, sale of hot dogs, and so on). Another reason for the team's profitability is that it owns the TV cable company that broadcasts the team's games and other events.

Assume that it is the end of 2004. The baseball team is for sale, and you have a wealthy client who is interested in buying the team. She thinks that she must offer $600 million to win the bidding for team ownership. She would borrow all of the money from a large regional bank. Although the team is expected to be profitable, it is not clear that the expected profits would be enough to allow the new owner to pay off the bank debt in a reasonable amount of time.

Newly passed anti-trust regulations in your state may require the new owner to sell the cable company after purchasing the team. However, the laws are unclear. Your client's lawyers say that they could get the courts to let her keep the cable company. She must decide whether she would want to try to keep the cable company or sell it. To keep it, she would have to fight for the cable company in court, which means hefty legal expenses.

The team's expected sources of revenue and expenses are discussed in the next sections.

Sources of Revenue

Revenue comes from the following sources, which will be discussed next:

- Selling tickets to regular-season baseball games
- Concession sales
- Playoff-game tickets and concession sales
- Revenue from "national" sources
- Cable TV revenue

Regular-Season Tickets and Concession Sales

The seating capacity of the baseball park is 34,000. There are 81 regular season games. The average ticket price in the 2004 season was $30. This average ticket price is expected to be 5% higher in each succeeding year through 2008. In a typical season, the team can, on average, be expected to sell 90% of seating capacity. However, in years that the team is very good, average ticket sales have been 95% of seating capacity.

Concession sales are for parking, beer, hot dogs, souvenirs, and so on. The team hires a company to run the concessions. The concession company pays a percentage of concession revenue to the team. There has been a fairly stable relationship between concession revenue and ticket sales. Concession revenue has typically been 30% of baseball ticket sales in the regular season.

Playoff-Game Tickets and Concession Sales

Teams that do well in the regular season are allowed to enter a post-season "playoff" period. Winning teams keep advancing in the playoffs. Eventually, the teams that progress to the end of the playoffs meet in the World Series. Capacity would be 100% at each playoff game. Ticket prices are much higher in the playoffs. The concession company raises its prices during the playoffs, and the 30% concession revenue ratio would apply in the playoffs also.

Case 7

Revenue From "National" Sources

The team is part of Major League Baseball, which has contracts with national television networks, such as ABC, CBS, and ESPN, and other national ventures. The teams share the revenue from these contracts. National revenue is the expected revenue each year from Major League Baseball contracts. This amount is expected to be $25 million in 2005 and to increase in later years.

Cable TV Revenue

If the cable company is kept, media revenue for 2005 is expected to be $50 million, increasing in succeeding years. Of course, if the cable company is sold, cable company operating revenue would be zero after the sale. (It would not actually change hands until 2006, so that the team would get the 2005 operating revenue in either case).

If the cable company is actually sold, revenue from *selling* the cable company would be $80 million in each year from 2006–2008. If the cable company is not sold, of course, this revenue would be zero. Revenue from selling the cable company would be zero in 2005 because the company would not change ownership until 2006.

Sources of Expenses

Expenses come from the following sources, which are described next:

- Player salaries
- Team operating expenses
- Cable company operating expenses
- Legal fees

Player Salaries

The primary operating expense for a baseball team is player salaries. The team's player compensation for 2004 was $110,000,000. Each year in Major League Baseball, there are good players who leave their team to be hired by other teams. These players are called "free agents." A team's management can control salaries to some extent. To reduce salaries, players whose contracts have expired are not rehired; they are allowed to become free agents, and they are replaced by lower-priced younger players. Or, management can hire talented free agents to improve the team. Of course, this increases salaries because good players cost more than mediocre or untested players. Your client tells you that a "low" salary structure for the hometown team would be $100 million per year; a "medium" salary structure would be $120 million per year; an "aggressive" salary structure would be $130 million per year. Salary structure has an impact on the team's success in playoff games in the following ways:

- Good players win more games than mediocre players.
- Making it into the playoffs is a function of having good players.
- Having good players is a function of the amount of money that management spends on salaries.

Of course, it's difficult to predict whether a professional baseball team will be successful in any given year. However, for purposes of the financial projection, your client must make some assumptions about the relationship of salary levels and games won. Thus, the number of expected playoff games is assumed to be based on the salary structure. With a "low" structure, zero playoff games per year are expected. With a "medium" structure, 9 playoff games per year are expected. With an "aggressive" structure, 12 playoff games per year are expected.

Team Operating Expenses

Other operating expenses are the rest of the team expenses—executive salaries, stadium upkeep, and so on. This amount is expected to be $55 million in 2005 and to increase somewhat each year thereafter. (This number does not include the cost of running the cable company.)

Cable Company Operating Expenses

The cost of operating the cable company is expected to be $13 million in 2005 and to increase somewhat each year thereafter. Of course, the operating cost would be zero if the cable company is not owned.

Legal Fees

If the new owner decided to try to keep the cable company, she will have $5 million in legal fees in both 2005 and in 2006. If she decided to sell the cable company, she would have zero legal fees—there would be no dispute to litigate with the state. In any case, legal fees for 2007 and 2008 would be zero.

Assignment 1 Creating a Spreadsheet for Decision Support

In this assignment, you will produce a spreadsheet that models the business decision. Then, in Assignment 2, you will use the spreadsheet to gather data and then write a memorandum to the prospective team owner that explains your recommended action. In addition, in Assignment 3, you will be asked to prepare an oral presentation of your recommendations.

Next, you will create the spreadsheet model of the baseball team purchase decision. The model is an income statement and cash flow forecast for the years 2005 to 2008. You will be given some hints on how each section should be set up before entering cell formulas. Your spreadsheet should have the sections that follow.

- **CONSTANTS**
- **INPUTS**
- **SUMMARY OF KEY RESULTS**
- **CALCULATIONS**
- **INCOME STATEMENT AND CASH FLOW STATEMENT**
- **DEBT OWED**

A discussion of each spreadsheet section follows. *When you type in the spreadsheet skeleton, follow the order given here.*

CONSTANTS Section

Your spreadsheet should have the constants shown in Figure 7-1. An explanation of some of the line items follows the figure. The cells should be formatted as shown.

Case 7

	A	B	C	D	E	F
1	**BASEBALL TEAM PURCHASE DECISION**					
2						
3	**CONSTANTS**	**2004**	**2005**	**2006**	**2007**	**2008**
4	TAX RATE	**NA**	0.31	0.32	0.33	0.34
5	CAPACITY OF PARK	34000	34000	34000	34000	34000
6	NUMBER OF REGULAR GAMES	81	81	81	81	81
7	TICKET PRICE - AVG	30	**NA**	**NA**	**NA**	**NA**
8	EXPECTED MEDIA REVENUE	46000000	50000000	56000000	69000000	79000000
9	INTEREST RATE	0.10	0.10	0.10	0.10	0.10
10	NATIONAL REVENUE	**NA**	25000000	27000000	30000000	33000000
11	PLAYOFF TICKET PRICE - AVG	**NA**	75	100	125	150
12	PLAYER COMPENSATION	110000000	**NA**	**NA**	**NA**	**NA**
13	CONCESSION REVENUE FACTOR	**NA**	0.30	0.30	0.30	0.30
14	MINIMUM CASH NEEDED TO START YEAR	**NA**	10000000	10000000	10000000	10000000
15	OTHER OPERATING EXPENSES	**NA**	55000000	57000000	59000000	62000000
16	CABLE COMPANY OPERATING EXPENSES	**NA**	13000000	14000000	15000000	16000000
17	TICKET PRICE INCREASE %	**NA**	0.05	**NA**	**NA**	**NA**

Figure 7-1 CONSTANTS section

- The TAX RATE expected in the years is expected to increase slightly, as you can see.
- The average regular-season game ticket price was $30 in 2004. The average price is expected to increase 5% each year. Thus, 2005's price will be 5% more than 2004's average price, 2006's price will be 5% more than 2005's price, and so on.
- The INTEREST RATE on debt owed to the bank in each of the four years is expected to be 10%.
- The team would need $10 million in cash to start each year. This is called the MINIMUM CASH NEEDED TO START YEAR.

INPUTS Section

Your spreadsheet should have the inputs shown in Figure 7-2. An explanation of the line items follows the figure.

	A	B	C	D	E	F
19	**INPUTS**	**2004**	**2005**	**2006**	**2007**	**2008**
20	CAPACITY % -- G=GOOD, H=HIGH	**NA**				
21	OWN MEDIA -- K=KEEP,S=SELL	**NA**		**NA**	**NA**	**NA**
22	SALARY STRATEGY -- L=LOW, M=MEDIUM, A=AGGRESIVE	**NA**		**NA**	**NA**	**NA**

Figure 7-2 INPUTS section

- CAPACITY % is the seating utilization. The team can be expected to sell 90% of their seats in a typical year. This would be considered "Good" capacity utilization. However, in some years, average ticket sales have been 95% of capacity, which is considered "High." You would enter a **G** or an **H**, depending on expected utilization in a year. The same entry could be made for all years, or it could be a mix of entries (e.g., a pattern of **GGHH** would be possible in the four years).
- OWN MEDIA refers to whether the owner should keep the cable company. Newly passed state anti-trust regulations may require the new owner to sell the cable company after the purchase of the baseball team. She thinks that her lawyers could get the courts to let her keep the cable company, however. She must decide whether she would want to try to keep the cable company or to sell it—you will enter a **K** or **S** to

indicate "Keep" or "Sell." The input applies to all years —a cell with **NA** should have no entry, and its address should not be referred to in any formula in the spreadsheet.

- SALARY STRATEGY refers to the salary structure strategy for player payment. Each year in Major League Baseball, there are good players that can be hired. Good players win more games than mediocre players, but good players cost more. There are three compensation levels: low, medium, and aggressive. Enter **L**, **M**, or **A** to indicate "Low," "Medium," or "Aggressive." The input applies to all years.

Your instructor may tell you to apply Conditional Formatting to the input cells, so out-of-bounds values are highlighted in some way (for example, in red type and/or in bold-face type). If so, use the Excel Help system to determine how to use Conditional Formatting for this purpose, or follow the instructions provided by your instructor. (*Hint*: For each input cell, you will need a formula that uses the Excel =AND( ) function.)

SUMMARY OF KEY RESULTS Section

Your spreadsheet should resemble Figure 7-3.

	A	B	C	D	E	F
25	**SUMMARY OF KEY RESULTS**	**2004**	**2005**	**2006**	**2007**	**2008**
26	NET INCOME TO TOTAL REVENUE RATIO	**NA**				
27	END-OF-THE-YEAR DEBT OWED	**NA**				

Figure 7-3 SUMMARY OF KEY RESULTS section

For each year, your spreadsheet should show (1) the ratio of net income after taxes to total revenue, which indicates how profitable the team is; and (2) the amount owed to the bank at year-end. The net income to total revenue ratio is a calculation and is echoed to this section. The amount owed at year-end is computed in the spreadsheet body and is echoed to this section. The ratio cells should be formatted for three decimal places.

CALCULATIONS Section

You should calculate various intermediate results, which are then used in the **INCOME STATEMENT AND CASH FLOW STATEMENT** section. Calculations are based on input values or on year 2004 values. When called for, use absolute addressing.

INCOME STATEMENT AND CASH FLOW STATEMENT Section

The forecast for net income and cash flow starts with the cash on hand at the beginning of the year. This is followed by the income statement and concludes with the calculation of cash on hand at year-end. Results in this section should be formatted for zero decimals (i.e., no pennies).

Here are some things to think about when structuring your income statement:

- The income statement should show revenues from all sources, followed by expenses from all sources.
- INCOME BEFORE INTEREST AND TAXES equals all revenues less all expenses, except for interest and taxes.
- INTEREST EXPENSE is based on the year's interest rate (a constant) and beginning-of-the-year debt owed.
- INCOME BEFORE TAXES equals income before interest and taxes less interest expense.

- INCOME TAX EXPENSE is zero if income before taxes is zero or less; otherwise, apply the tax rate to income before taxes.
- NET INCOME AFTER TAXES equals income before taxes less income tax expense.

Continuing with this statement, line items for the year-end cash calculation are discussed. In Figure 7-4, column B is for year 2004, column C for year 2005, and so forth.

	A	B	C	D	E	F
59	OUTLAY FOR TEAM	NA	300000000	300000000	0	0
60						
61	NET CASH POSITION (NCP) BEFORE BORROWING AND REPAYMENT OF DEBT (BEGINNING-OF-THE-YEAR CASH PLUS NET INCOME AFTER TAXES)	NA				
62	ADD: BORROWING FROM BANK	NA				
63	LESS: REPAYMENT TO BANK	NA				
64	EQUALS: END-OF-THE-YEAR CASH ON HAND	30000000				

Figure 7-4 CASH ON HAND calculation

- Year 2004 values (column B) are mostly NA, except $30 million END-OF-THE-YEAR CASH ON HAND.
- THE NET CASH POSITION (NCP) at the end of a year equals the cash at the beginning of a year, plus the year's net income after taxes, less the outlay for the team. Assume there are no receivables or payables. The outlay for the team would be $300 million in each year 2005 and 2006.
- Your client's bank will lend her enough money at year-end to get to the minimum needed cash. If the NCP is less than minimum cash, then she must borrow enough to reach the minimum cash.
- If the NCP is more than the minimum cash at the end of a year and there is outstanding debt, then as much of the debt as possible should be repaid (but not to take the team below the minimum cash).
- END-OF-THE-YEAR CASH ON HAND equals the NCP, plus any borrowing, less any repayment.

DEBT OWED Section

Your spreadsheet body ends with a calculation of debt owed at year-end. You can assume that the team owed $10 million at the end of 2004, and the new owner would take over this debt.

Assignment 2 Using the Spreadsheet for Decision Support

You will now complete the case by (1) using the spreadsheet to gather the data needed to make the purchase decision, and (2) documenting your recommendation in a memorandum.

Assignment 2A: Using the Spreadsheet to Gather Data

You have built the spreadsheet to model the purchase decision. Your client says that High capacity at regular season games would be likely only if the team was very good, which implies an Aggressive salary structure. Otherwise, merely Good capacity utilization is likely to occur. In either of those cases, the cable company could be owned or not owned. You are told that the client and the bank want to know about the profitability and debt owed in each of the following six scenarios:

- Good capacity, Keep cable company, Low salary structure
- Good capacity, Keep cable company, Medium salary structure
- Good capacity, Sell cable company, Low salary structure
- Good capacity, Sell cable company, Medium salary structure
- High capacity, Keep cable company, Aggressive salary structure
- High capacity, Sell cable company, Aggressive salary structure

Now run "what if" scenarios with the possible six sets of input values. The way you do this depends on whether your instructor has told you to use the Scenario Manager and/or whether you have been told to chart any output values.

- If your instructor has told you *not* to use the Scenario Manager, you must now manually enter the input value combinations. Note the results for each on a piece of paper. (The results will show in the Summary of Key Results as you work.) Additionally, if your instructor told you to chart the 2007–2008 results, manually enter these into a separate spreadsheet section beneath the DEBT OWED section. (You will need room for each of the six alternatives for 2007 and 2008.)
- If your instructor has told you to use the Scenario Manager, perform the procedures set forth in Tutorial C to set up and run the Scenario Manager. Record the six possible scenarios. The changing cells are the cells used to input the capacity, cable ownership, and salary structure values. Output cells are for 2007 and 2008 only. (In the Scenario Manager, you can enter noncontiguous input ranges as follows: C20..F20, C21, C22.) If your instructor told you to chart the 2007–2008 results, you can chart the Summary Sheet values, which are nicely arrayed for that purpose.
- In either case, when you are done gathering data, print the entire workbook (including any charts on their own sheets). Then save the **.xls** file. **BBALL.xls** is recommended as the file name.

Assignment 2B: Documenting Your Recommendations in a Memorandum

Open MS Word and write a brief memorandum to your client, who wants to know which purchase strategy seems best. Her goals are to have a net-income-to-revenue ratio exceeding 5% in each year 2007 and 2008 *and* to have less than $500 million in debt at the end of 2008. She would be reluctant to buy the team if there were no strategies that would achieve those financial goals.

- Your memorandum should have a proper heading (DATE / TO / FROM / SUBJECT). You may wish to use a Word memo template (File—New, click Memos, choose Contemporary Memo, and click OK).
- In the first paragraph, tell your client which Capacity/Own Media/Salary Structure strategies will achieve her goals, if any. In addition, you should point out the one that seems to be optimal and explain your reasoning.

- Support your statement graphically as your instructor requires: (1) If you used the Scenario Manager, go back to Excel and put a copy of the Summary Sheet results into the Windows Clipboard. Then, in Word, copy this from the Clipboard. (Tutorial C refers to this procedure.) (2) If you made a chart in Excel, copy it from Excel to Word in the same way. (3) Make a summary table in Word and position it after the memo's first paragraph. This procedure is described next.

Enter a table into Word, using the following procedure:

1. Select the **Table** menu option, click **Insert**, and then click **Table**.
2. Enter the number of rows and columns.
3. Select **AutoFormat** and choose **Table Grid 1**.
4. Select **OK**, and then select **OK** again.

Your table should resemble the format of the table shown in Figure 7-5.

Capacity	Cable Company	Salary Structure	2007 Ratio	2008 Ratio	2008 Year-end Debt
Good	Keep	Low			
Good	Keep	Medium			
Good	Sell	Low			
Good	Sell	Medium			
High	Keep	Aggressive			
High	Sell	Aggressive			

Figure 7-5 Format of table to insert in memorandum

Assignment 3 Giving an Oral Presentation

Your instructor may request that you also present your analysis and recommendations in an oral presentation. If so, assume that the client has accepted the recommendation that you set forth in your memo. She has asked you to give a presentation explaining your recommendation to a joint meeting of the team management and the bankers who would hold promissory notes for debt owed. Prepare to explain your analysis and recommendation to the group in 10 minutes or less. Use visual aids or handouts that you think are appropriate. Tutorial E in this casebook has guidance on how to prepare an oral presentation.

Deliverables

Assemble the following deliverables for your instructor:

1. Memorandum
2. Spreadsheet printouts
3. Diskette, which should have your Word memo file and your Excel spreadsheet file.

Staple the printouts together, with the memo on top. If there is more than one **.xls** file on the diskette, write your instructor a note, stating the name of your model's **.xls** file.

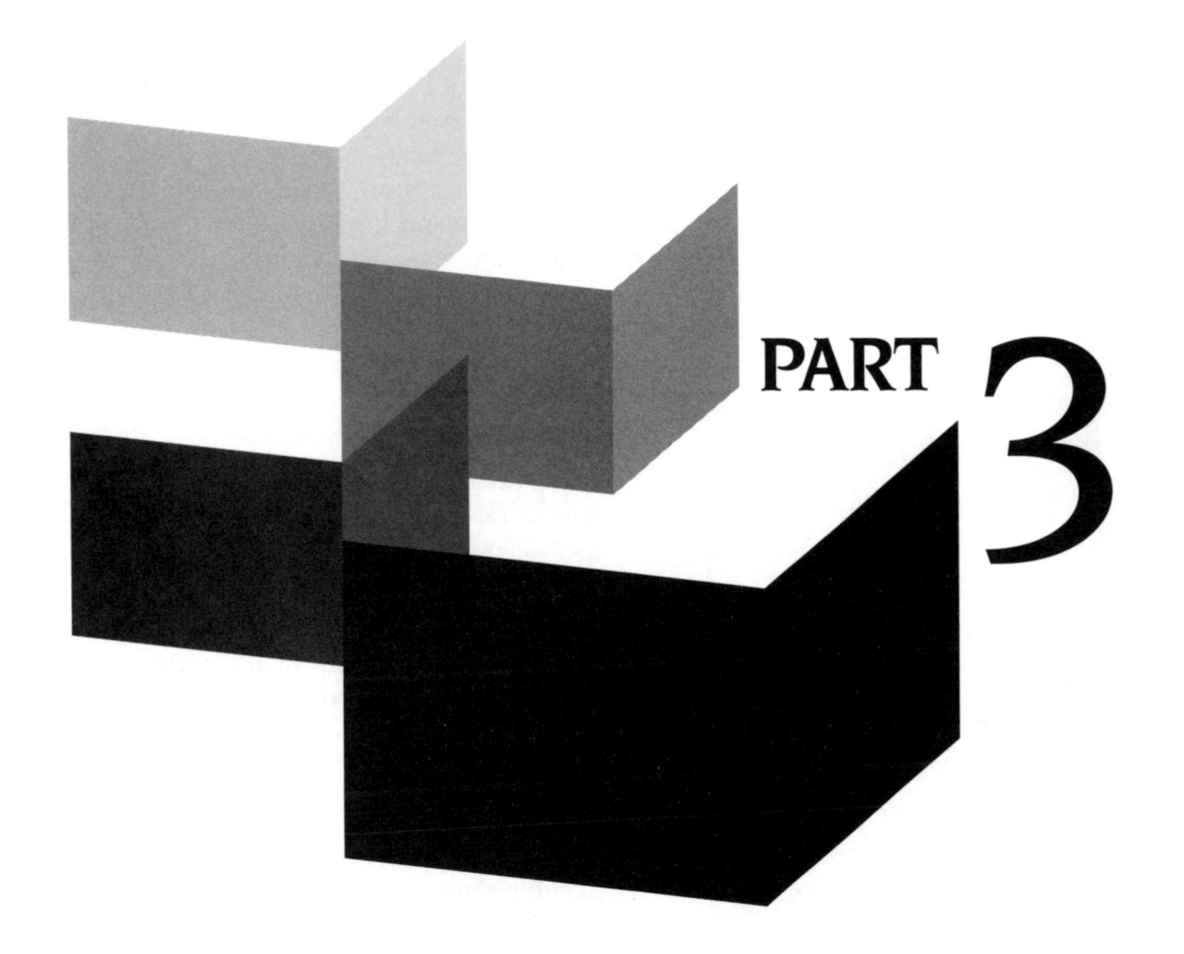

Decision Support Cases Using the Excel Solver

Building a Decision Support System Using the Excel Solver

Decision Support Systems (DSS) help people to make decisions. (The nature of DSS programs is discussed in Tutorial C.) Tutorial D teaches you how to use the Solver, one of Excel's built-in decision support tools.

For some business problems, decision makers want to know the best, or optimal, solution. Usually this means maximizing a variable (e.g., net income) or minimizing another variable (e.g., total costs). This optimization is subject to constraints, which are rules that must be observed when solving a problem. The Solver computes answers to such optimization problems.

This tutorial has four sections:

1. **Using the Excel Solver** In this section, you'll learn how to use the Solver in decision-making. As an example, you use the Solver to create a production schedule for a sporting goods company. This schedule is called the Base Case.
2. **Extending the Example** In this section, you'll test what you've learned about using the Solver as you modify the sporting goods company's production schedule. This is called the Extension Case.
3. **Using the Solver on a New Problem** In this section, you'll use the Solver on a new problem.
4. **Trouble-shooting the Solver** In this section, you'll learn how to overcome problems you might encounter when using the Solver.

Tutorial C has some guidance on basic Excel concepts, such as formatting cells and using functions, such as =IF(). Refer to Tutorial C for a review of such topics.

Using the Excel Solver

Suppose that a company must set a production schedule for its various products, each of which has a different profit margin (selling price less costs). At first, you might assume that the company will maximize production of all profitable products to maximize net income. However, a company typically cannot make and sell an unlimited number of its products because of constraints.

One constraint affecting production is the "shared resource problem." For example, several products in a manufacturer's line might require the same raw materials, which are in limited supply. Similarly, the manufacturer might require the same machines to make several of its products. In addition, there might also be a limited pool of skilled workers available to make the products.

In addition to production constraints, sometimes management's policies impose constraints. For example, management might decide that the company must have a broader product line. As a consequence, a certain production quota for several products must be met, regardless of profit margins.

Thus, management must find a production schedule that will maximize profit, given the constraints. Optimization programs like the Solver look at each combination of products, one after the other, ranking each combination by profitability. Then the program reports the most profitable combination.

To use the Solver, you'll set up a model of the problem, including the factors that can vary, the constraints on how much they can vary, and the goal you are trying to maximize (usually net income) or minimize (usually total costs). The Solver then computes the best solution.

Setting Up a Spreadsheet Skeleton

Suppose that your company makes two sporting goods products—basketballs and footballs. Assume that you will sell all the balls you produce. To maximize net income, you want to know how many of each kind of ball to make in the coming year.

Making each kind of ball requires a certain (and different) number of hours, and each ball has a different raw materials cost. Because you have only a limited number of workers and machines, you can devote a maximum of 40,000 hours to production. This is a shared resource. You do not want that resource to be idle, however. Downtime should be no more than 1,000 hours in the year, so machines should be used for at least 39,000 hours.

Marketing executives say you cannot make more than 60,000 basketballs and may not make less than 30,000. Furthermore, you must make at least 20,000 footballs but not more than 40,000. Marketing says the ratio of basketballs to footballs produced should be between 1.5 and 1.7—i.e., more basketballs than footballs, but within limits.

What would be the best production plan? This problem has been set up in the Solver. The spreadsheet sections are discussed in the pages that follow.

AT THE KEYBOARD

Start out by saving the blank spreadsheet as **SPORTS1.xls**. Then you should enter the skeleton and formulas as they are discussed.

CHANGING CELLS Section

The **CHANGING CELLS** section contains the variables the Solver is allowed to change while it looks for the solution to the problem. Figure D-1 shows the skeleton of this spreadsheet section and the values that you should enter. An analysis of the line items follows the figure.

	A	B
1	**SPORTING GOODS EXAMPLE**	
2	CHANGING CELLS	
3	NUMBER OF BASKETBALLS	1
4	NUMBER OF FOOTBALLS	1

Figure D-1 CHANGING CELLS section

- The changing cells are for the number of basketballs and footballs to be made and sold. The changing cells are like input cells, except Solver (not you) plays "what-if" with the values, trying to maximize or minimize some value (in this case, maximize net income).
- Note that some number should be put in the changing cells each time before the Solver is run. It's customary to put the number 1 into the changing cells (as shown). Solver will change these values when the program is run.

CONSTANTS Section

Your spreadsheet should also have a section for values that will not change. Figure D-2 shows a skeleton of the **CONSTANTS** section and the values you should enter. A discussion of the line items follows the figure.

You should use Format— Cells—Number to set the constants range to two decimal places.

	A	B
6	**CONSTANTS**	
7	BASKETBALL SELLING PRICE	14.00
8	FOOTBALL SELLING PRICE	11.00
9	TAX RATE	0.28
10	NUMBER OF HOURS TO MAKE A BASKETBALL	0.50
11	NUMBER OF HOURS TO MAKE A FOOTBALL	0.30
12	COST OF LABOR -- 1 MACHINE HOUR	10.00
13	COST OF MATERIALS -- 1 BASKETBALL	2.00
14	COST OF MATERIALS -- 1 FOOTBALL	1.25

Figure D-2 CONSTANTS section

- The SELLING PRICE for one basketball and for one football is shown.
- The TAX RATE is the rate applied to income before taxes to compute income tax expense.
- The NUMBER OF MACHINE HOURS needed to make a basketball and a football is shown. Note that a ball-making machine can produce two basketballs in an hour.
- COST OF LABOR: A ball is made by a worker using a ball-making machine. A worker is paid $10 for each hour he or she works at a machine.
- COST OF MATERIALS: The costs of raw materials for a basketball and football are shown.

Notice that the profit margins (selling price less costs of labor and materials) for the two products are not the same. They have different selling prices and different inputs (raw materials, hours to make)—and the inputs have different costs per unit. Also note that you cannot tell from the data how many hours of the shared resource (machine hours) will be devoted to basketballs and how many to footballs, because you don't know in advance how many basketballs and footballs will be made.

CALCULATIONS Section

In the **CALCULATIONS** section, you will calculate intermediate results that (1) will be used in the spreadsheet body, and/or (2) will be used as constraints. First, use Format—Cells—Number to set the calculations range to two decimal places. Figure D-3 shows the skeleton and formulas that you should enter. A discussion of the cell formulas follows the figure.

NOTE

Cell widths are changed here merely to show the formulas—you need not change the width.

	A	B
16	CALCULATIONS	
17	RATIO OF BASKETBALLS TO FOOTBALLS	=B3/B4
18	TOTAL BASKETBALL HOURS USED	=B3*B10
19	TOTAL FOOTBALL HOURS USED	=B4*B11
20	TOTAL MACHINE HOURS USED (BB + FB)	=B18+B19

Figure D-3 CALCULATIONS section cell formulas

- The RATIO OF BASKETBALLS TO FOOTBALLS (cell B17) will be needed in a constraint.
- TOTAL BASKETBALL HOURS USED: The number of machine hours needed to make all basketballs (B3 * B10) is computed in cell B18. Cell B10 has the constant for the hours needed to make one basketball. Cell B3 (a changing cell) has the number of basketballs made. (Currently, this cell shows one ball, but that number will change when the Solver works on the problem.)
- TOTAL FOOTBALL HOURS USED: The number of machine hours needed to make all footballs is calculated similarly, in cell B19.
- TOTAL MACHINE HOURS USED (BB + FB): The number of hours needed to make both kinds of balls (cell B20) will be a constraint; this value is the sum of the hours just calculated for footballs and basketballs.

Notice that constants in the Excel cell formulas in Figure D–3 are referred to by their cell addresses. Use the cell address of a constant rather than hard-coding a number in the Excel expression: If the number must be changed later, you only have to change it in the CONSTANTS section cell, not in every cell formula in which you used the value.

Notice that you do not calculate the amounts in the changing cells (here, the number of basketballs and footballs to produce). The Solver will compute those numbers. Also notice that you can use the changing cell addresses in your formulas. When you do that, you assume the Solver has put the optimal values in each changing cell; your expression makes use of that number.

Figure D-4 shows the values after Excel evaluates the cell formulas (with 1's in the changing cells):

	A	B
16	**CALCULATIONS**	
17	RATIO OF BASKETBALLS TO FOOTBALLS	1.00
18	TOTAL BASKETBALL HOURS USED	0.50
19	TOTAL FOOTBALL HOURS USED	0.30
20	TOTAL MACHINE HOURS USED (BB + FB)	0.80

Figure D-4 CALCULATIONS section cell values

INCOME STATEMENT Section

The target value is calculated in the spreadsheet body in the INCOME STATEMENT section. This is the value that the Solver is expected to maximize or minimize. The spreadsheet body can take any form. In this textbook's Solver cases, the spreadsheet body will be an income statement. Figure D-5 shows the skeleton and formulas that you should enter. A discussion of the line-item cell formulas follows the figure.

Income statement cells were formatted for two decimal places.

	A	B
22	**INCOME STATEMENT**	
23	BASKETBALL REVENUE (SALES)	=B3*B7
24	FOOTBALL REVENUE (SALES)	=B4*B8
25	TOTAL REVENUE	=B23+B24
26	BASKETBALL MATERIALS COST	=B3*B13
27	FOOTBALL MATERIALS COST	=B4*B14
28	COST OF MACHINE LABOR	=B20*B12
29	TOTAL COST OF GOODS SOLD	=SUM(B26:B28)
30	INCOME BEFORE TAXES	=B25-B29
31	INCOME TAX EXPENSE	=IF(B30<=0,0,B30*B9)
32	NET INCOME AFTER TAXES	=B30-B31

Figure D-5 INCOME STATEMENT section cell formulas

- REVENUE (cells B23 and B24) equals the number of balls times the respective unit selling price. The number of balls is in the changing cells, and the selling prices are constants.
- MATERIALS COST (cells B26 and B27) follows a similar logic: number of units times unit cost.
- The COST OF MACHINE LABOR is the calculated number of machine hours times the hourly labor rate for machine workers.
- TOTAL COST OF GOODS SOLD is the sum of the cost of materials and the cost of labor.

- This is the logic of income tax expense: If INCOME BEFORE TAXES is less than or equal to zero, the tax is zero; otherwise, the income tax expense equals the tax rate times income before taxes. An =IF() statement is needed in cell B31.

Excel evaluates the formulas. Figure D-6 shows the results (assuming 1s in the changing cells):

	A	B
22	**INCOME STATEMENT**	
23	BASKETBALL REVENUE (SALES)	14.00
24	FOOTBALL REVENUE (SALES)	11.00
25	TOTAL REVENUE	25.00
26	BASKETBALL MATERIALS COST	2.00
27	FOOTBALL MATERIALS COST	1.25
28	COST OF MACHINE LABOR	8.00
29	TOTAL COST OF GOODS SOLD	11.25
30	INCOME BEFORE TAXES	13.75
31	INCOME TAX EXPENSE	3.85
32	NET INCOME AFTER TAXES	9.90

Figure D-6 INCOME STATEMENT section cell values

Tutorial D

Constraints

Constraints are rules which the Solver must observe when computing the optimal answer to a problem. Constraints will need to refer to calculated values, or to values in the spreadsheet body. Therefore, you must build those calculations into the spreadsheet design, so they are available to your constraint expressions. (There is no section on the face of the spreadsheet for constraints. You'll use a separate window to enter constraints.)

Figure D-7 shows the English and Excel expressions for the basketball and football production problem constraints. A discussion of the constraints follows the figure.

Expression in English	Excel Expression
TOTAL MACHINE HOURS >= 39000	B20 >= 39000
TOTAL MACHINE HOURS <= 40000	B20 <= 40000
MIN BASKETBALLS = 30000	B3 >= 30000
MAX BASKETBALLS = 60000	B3 <= 60000
MIN FOOTBALLS = 20000	B4 >= 20000
MAX FOOTBALLS = 40000	B4 <= 40000
RATIO BB'S TO FB'S-MIN = 1.5	B17 >= 1.5
RATIO BB'S TO FB'S-MAX = 1.7	B17 <= 1.7
NET INCOME MUST BE POSITIVE	B32 >= 0

Figure D-7 Solver Constraint expressions

- As shown in Figure D-7, notice that a cell address in a constraint expression can be a changing cell address, a cell address in the constants section, a cell address in the calculations section, or a cell address in the spreadsheet body.
- You'll often need to set minimum and maximum boundaries for variables. For example, the number of basketballs (MIN and MAX) varies between 30,000 and 60,000 balls.
- Often, a boundary value is zero because you want the Solver to find a non-negative result. For example, here you want only answers that yield a positive net income. You tell the Solver that the amount in the net income cell must equal or exceed zero, so the Solver does not find an answer that produces a loss.
- Machine hours must be shared between the two kinds of balls. The constraints for the shared resource are: B20 >= 39,000 and B20 <= 40000, where cell B20 shows the total hours used to make both the basketballs and footballs. The shared-resource constraint seems to be the most difficult kind of constraint for students to master when learning the Solver.

Running the Solver: Mechanics

To set up the Solver, you must tell the Solver these things:

1. The cell address of the "target" variable that you are trying to maximize (or minimize, as the case may be)
2. The changing cell addresses
3. The expressions for the constraints

The Solver will put its answers in the changing cells and on a separate sheet.

Beginning to Set Up the Solver

AT THE KEYBOARD

To start setting up the Solver, select Tools—Solver. The first thing you will see is a Solver Parameters window, as shown in Figure D-8. Use the Solver Parameters window to specify the target cell, the changing cells, and the constraints. If you don't see the Solver tool under the Tools menu, you may need to activate it by going to Tools—Add-ins and clicking the Solver Add-in box to install it.

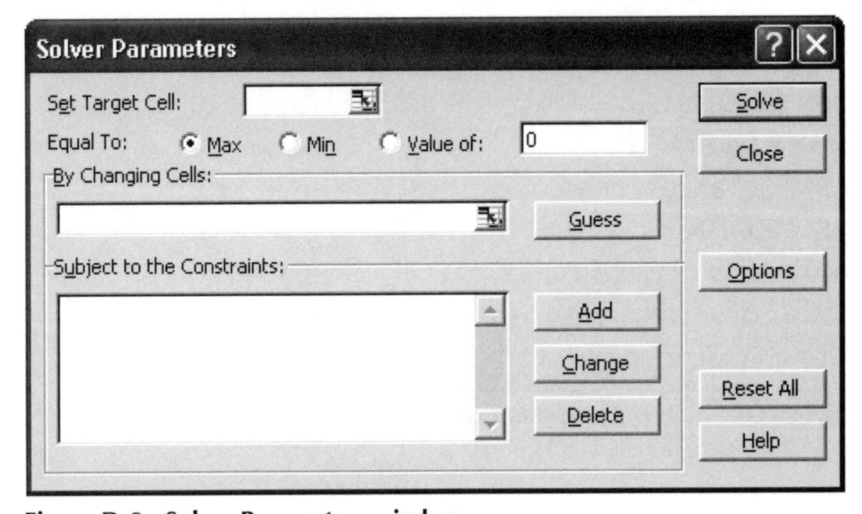

Figure D-8 Solver Parameters window

Setting the Target Cell

To set a target cell, use the following procedure:

1. The Target Cell is net income, cell B32.
2. To set the Target Cell, click in that input box and enter B32.
3. Max is the default; accept it here.
4. Enter a "0" for no desired net income value (Value of).

Figure D-9 shows entering data in the Target Cell.

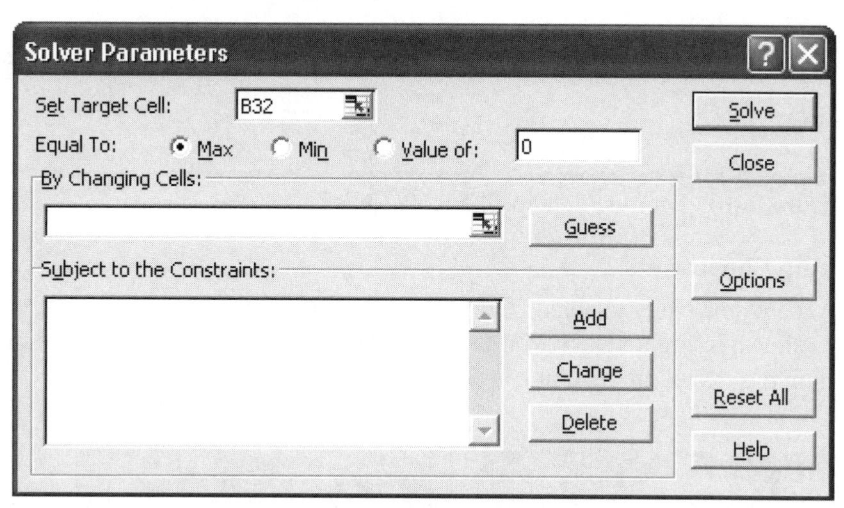

Figure D-9 Entering data in the Target Cell

Tutorial D

DO NOT hit Enter when you finish. You'll navigate within this window by clicking in the next input box.

When you enter the cell address, Solver may put in dollar signs, as if for absolute addressing. Ignore them—do not try to delete them.

Setting the Changing Cells

The changing cells are the cells for the balls, which are in the range of cells B3:B4. Click in the Changing Cells box and enter B3:B4, as shown in Figure D10. (Do *not* then hit Enter.)

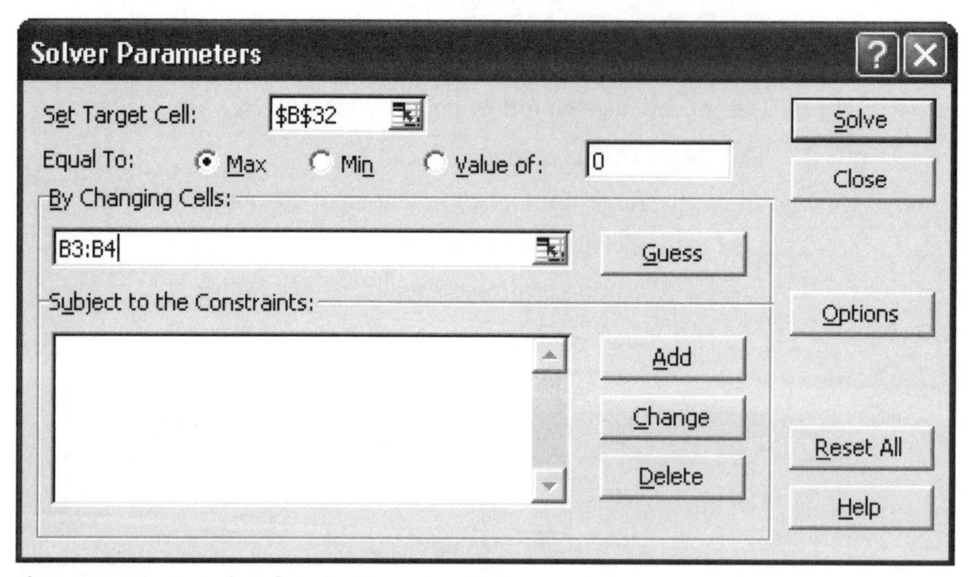

Figure D-10 Entering data in Changing Cells

Entering Constraints

You are now ready to enter the constraint formulas one by one. To start, click the Add button. As shown in Figure D-11, you'll see the Add Constraint window (here, shown with the minimum basketball production constraint entered).

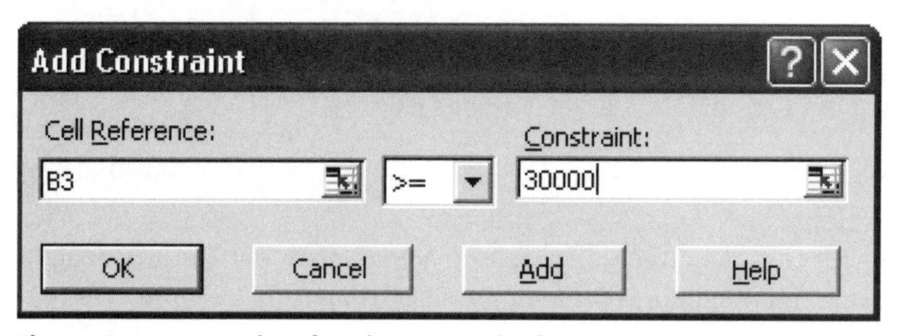

Figure D-11 Entering data in Constraint box

You should note the following about entering constraints and Figure D-11:

- To enter a constraint expression, do four things: (1) Type the variable's cell address in the left Cell Reference input box; (2) select the operator (<=, =, >=) in the smaller middle box; (3) enter the expression's right-hand side value, which is either a raw number or the cell address of a value, into the Constraint box; and (4) click Add to enter the constraint into the program. If you change your mind about the expression and do not want to enter it, click Cancel.
- The minimum basketballs constraint is: B3 >= 30000. Enter that constraint now. (Later, Solver may put an "equals" sign in front of the 30,000 and dollar signs in the cell reference.)
- After entering the constraint formula, click the Add button. This puts the constraint into the Solver model. It also leaves you in the Add Constraint window, allowing you to enter other constraints. You should enter those now. See Figure D-7 for the logic.
- When you're done entering constraints, click the Cancel button. This takes you back to the Solver Parameters window.

You should not put an expression into the Cell Reference window. For example, the constraint for the minimum basketball-to-football ratio is B3/B4 >= 1.5. You should not put =B3/B4 into the Cell Reference box. This is why the ratio is computed in the Calculations section of the spreadsheet (in cell B17). When adding that constraint, enter B17 in the Cell Reference box. (You are allowed to put an expression into the Constraint box, although that technique is not shown here and is not recommended.)

After entering all the constraints, you'll be back at the Solver Parameters window. You will see the constraints have been entered into the program. Not all constraints will show, due to the size of the box. The top part of the box's constraints area looks like the portion of the spreadsheet shown in Figure D-12.

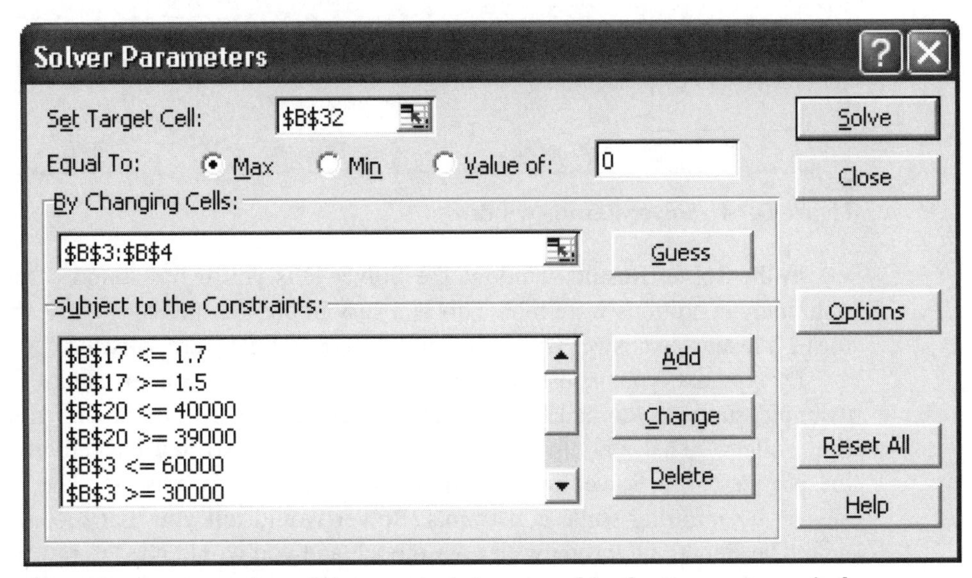

Figure D-12 A portion of the constraints entered in the Parameters window

Using the scroll arrow, reveal the rest of the constraints, as shown in Figure D-13.

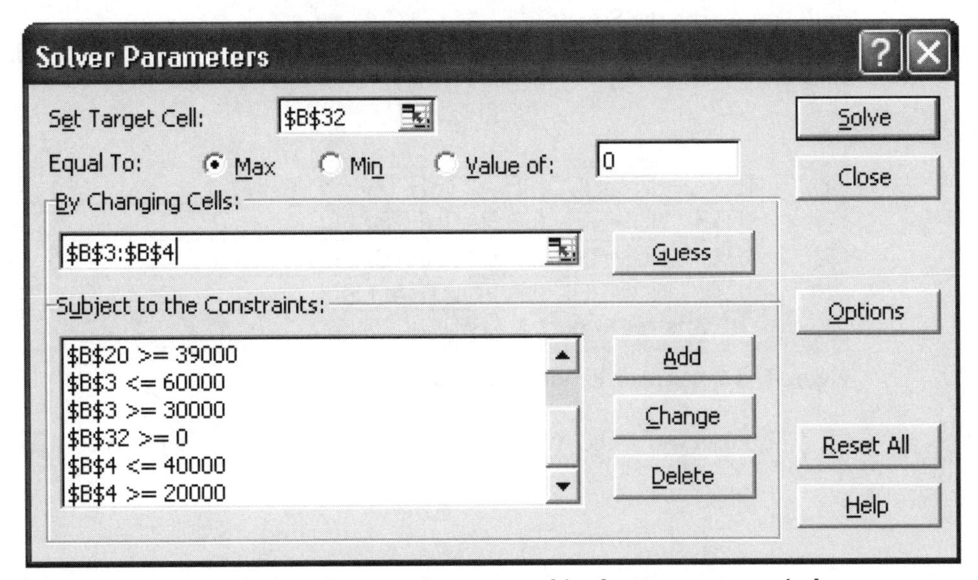

Figure D-13 Remainder of constraints entered in the Parameters window

Computing the Solver's Answer

To have the Solver actually calculate answers, click Solve in the upper-right corner of the Solver Parameters window. Solver does its work in the background—you do not see the internal calculations. Then the Solver gives you a Solver Results window, as shown in Figure D-14.

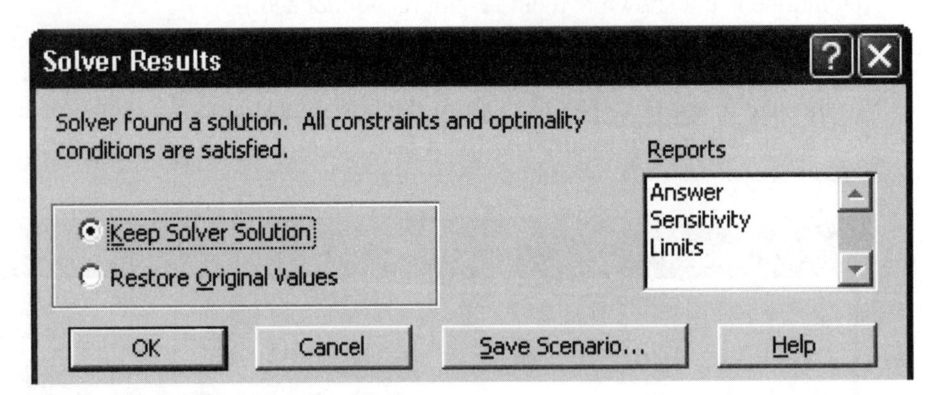

Figure D-14 Solver Results window

In the Solver Results window, the Solver tells you it has found a solution and that the optimality conditions were met. This is a very important message—you should always check for it. It means an answer was found and the constraints were satisfied.

By contrast, your constraints might be such that the Solver cannot find an answer. For example, suppose you had a constraint that said, in effect: "Net income must be at least a billion dollars." That amount cannot be reached, given so few basketballs and footballs and these prices. The Solver would report that no answer is feasible. The Solver may find an answer by ignoring some constraints. Solver would tell you that too. In either case, there would be something wrong with your model, and you would need to rework it.

There are two ways to see your answers. One way is to click OK. This lets you see the new changing cell values. A more formal (and complete) way is to click Answer in the Reports box, and then click OK. This puts detailed results into a new sheet in your Excel book. The new sheet is called an Answer Report. All answer reports are numbered sequentially as you run the Solver.

To see the Answer Report, click its tab, as shown in Figure D-15. (Here, this is Answer Report 1.)

23	B32	NET INCOME AFTER TAXES	473142.8
24	B3	NUMBER OF BASKETBALLS	57142.853
25	B3	NUMBER OF BASKETBALLS	57142.853
26	B4	NUMBER OF FOOTBALLS	38095.24

Answer Report 1 / Sheet1 / Sheet2 / Sheet3

Figure D-15 Answer Report Sheet tab

This takes you to the Answer Report. The top portion of the report is shown in Figure D-16.

	A	B	C	D	E
1	**Microsoft Excel 11.0 Answer Report**				
2	**Worksheet: [SPORTS1.XLS]Sheet1**				
3	**Report Created: 1/27/2004 12:01:55 PM**				
4					
5					
6	Target Cell (Max)				
7		**Cell**	**Name**	**Original Value**	**Final Value**
8		B32	NET INCOME AFTER TAXES	9.90	473142.87
9					
10					
11	Adjustable Cells				
12		**Cell**	**Name**	**Original Value**	**Final Value**
13		B3	NUMBER OF BASKETBALLS	1	57142.85348
14		B4	NUMBER OF FOOTBALLS	1	38095.2442

Figure D-16 Top portion of the Answer Report

Here is the remainder of the Answer Report, as shown in Figure D-17.

	A	B	C	D	E	F
17	Constraints					
18		**Cell**	**Name**	**Cell Value**	**Formula**	**Status**
19		B17	RATIO OF BASKETBALLS TO FOOTBALLS	1.50	B17<=1.7	Not Binding
20		B17	RATIO OF BASKETBALLS TO FOOTBALLS	1.50	B17>=1.5	Binding
21		B20	TOTAL MACHINE HOURS USED (BB + FB)	40000.00	B20<=40000	Binding
22		B20	TOTAL MACHINE HOURS USED (BB + FB)	40000.00	B20>=39000	Not Binding
23		B32	NET INCOME AFTER TAXES	473142.87	B32>=0	Not Binding
24		B3	NUMBER OF BASKETBALLS	57142.85348	B3>=30000	Not Binding
25		B3	NUMBER OF BASKETBALLS	57142.85348	B3<=60000	Not Binding
26		B4	NUMBER OF FOOTBALLS	38095.2442	B4>=20000	Not Binding
27		B4	NUMBER OF FOOTBALLS	38095.2442	B4<=40000	Not Binding

Figure D-17 Remainder of Answer Report

At the beginning of this tutorial, the changing cells had a value of 1, and the income was $9.90 (Original Value). The optimal solution values (Final Value) are also shown: $473,142.87 for net income (the target), and 57,142.85 basketballs and 38,095.24 footballs for the changing (adjustable) cells. (Of course you cannot make a part of a ball. The Solver can be asked to find only integer solutions; this technique is discussed at the end of this tutorial.)

The report also shows detail for the constraints: the constraint expression and the value that the variable has in the optimal solution. "Binding" means the final answer caused Solver to bump up against the constraint. For example, the maximum number of machine hours was 40,000, and that is the value Solver used in finding the answer.

"Not Binding" means the reverse. A better word for "binding" might be "constraining." For example, the 60,000 maximum basketball limit did not constrain the Solver.

The procedures used to change (edit) or delete a constraint are discussed later in this tutorial.

Print the worksheets (Answer Report and Sheet1). Save the Excel file (File—Save). Then, use File—Save As to make a new file called **SPORTS2.xls**, to be used in the next section of this tutorial.

EXTENDING THE EXAMPLE

Next, you'll modify the sporting goods spreadsheet. Suppose that management wants to know what net income would be if certain constraints were changed. In other words, management wants to play "what-if" with certain Base Case constraints. The resulting second case is called the Extension Case. Let's look at some changes to the original Base Case conditions.

- Assume that maximum production constraints will be removed.
- Similarly, the basketball-to-football production ratios (1.5, 1.7) will be removed.
- There will still be minimum production constraints at some low level: Assume that at least 30,000 basketballs and 30,000 footballs will be produced.
- The machine-hours shared resource imposes the same limits as it did previously.
- A more ambitious profit goal is desired: The ratio of net income after taxes to total revenue should be greater than or equal to .33. This constraint will replace the constraint calling for profits greater than zero.

AT THE KEYBOARD

Begin by putting 1s in the changing cells. You will need to compute the ratio of net income after taxes to total revenue. Enter that formula in cell B21. (The formula should have the net income after taxes cell address in the numerator and the total revenue cell address in the denominator.) In the extension case, the value of this ratio for the Solver's optimal answer must be at least .33. Click the Add button and enter that constraint.

Then, in the Solver Parameters window, constraints that are no longer needed are highlighted (select by clicking) and deleted (click the Delete button). Do that for the net income >= 0 constraint, the maximum football and basketball constraints, and the basketball-to-football ratio constraints.

The minimum football constraint must be modified, not deleted. Select that constraint, then click Change. That takes you to the Add Constraint window. Edit the constraint so 30,000 is the lower boundary.

When you are finished with the constraints, your Solver Parameters window should look like the one shown in Figure D-18.

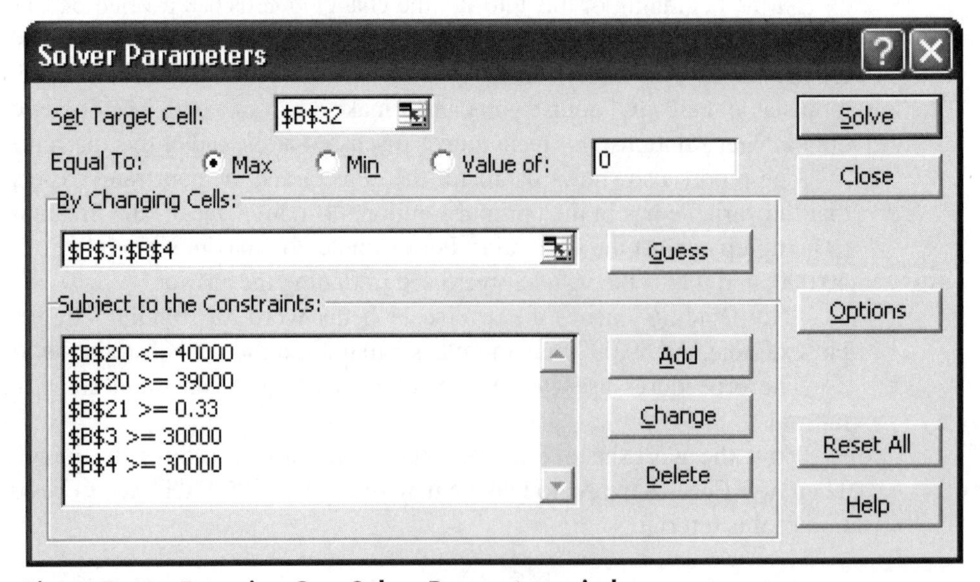

Figure D-18 Extension Case Solver Parameters window

You can tell Solver to solve for integer values. Here, cells B3 and B4 should be whole numbers. You use the Int constraint to do that. Figure D-19 shows entering the Int constraint.

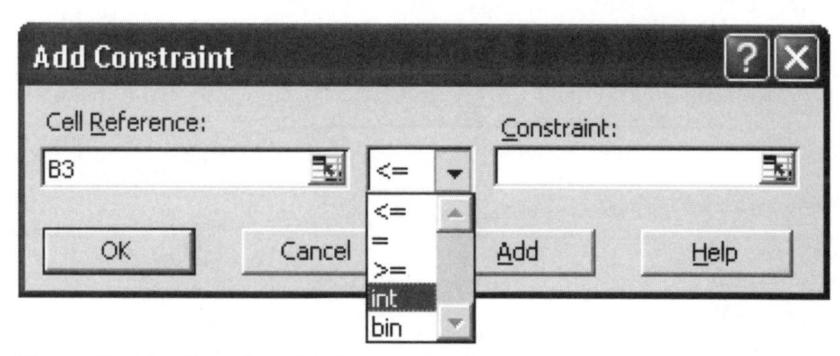

Figure D-19 Entering the Int constraint

Make those constraints for the changing cells. Your constraints should now look like the beginning portion of those shown in Figure D-20.

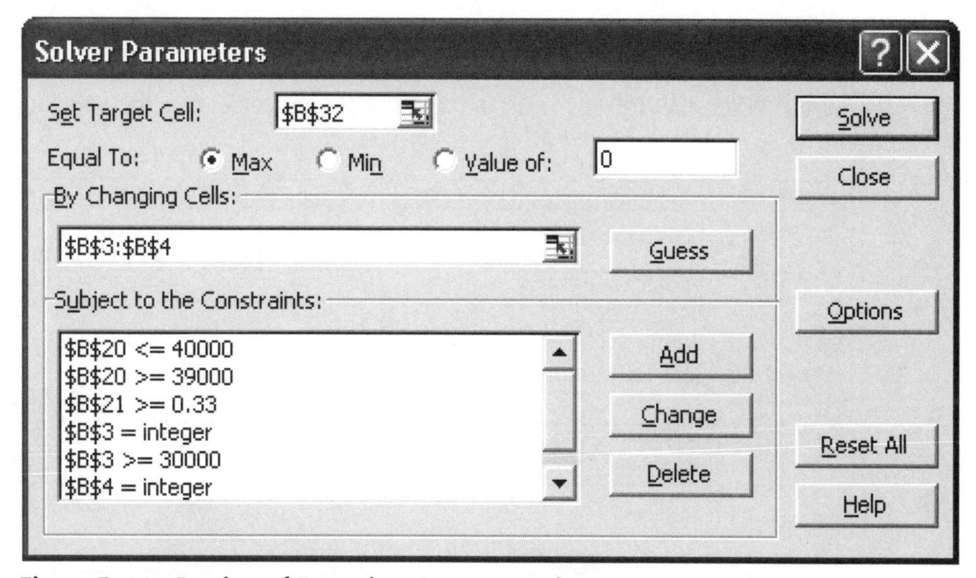

Figure D-20 Portion of Extension Case constraints

Scroll to see the remainder of the constraints, as shown in Figure D-21.

Tutorial D

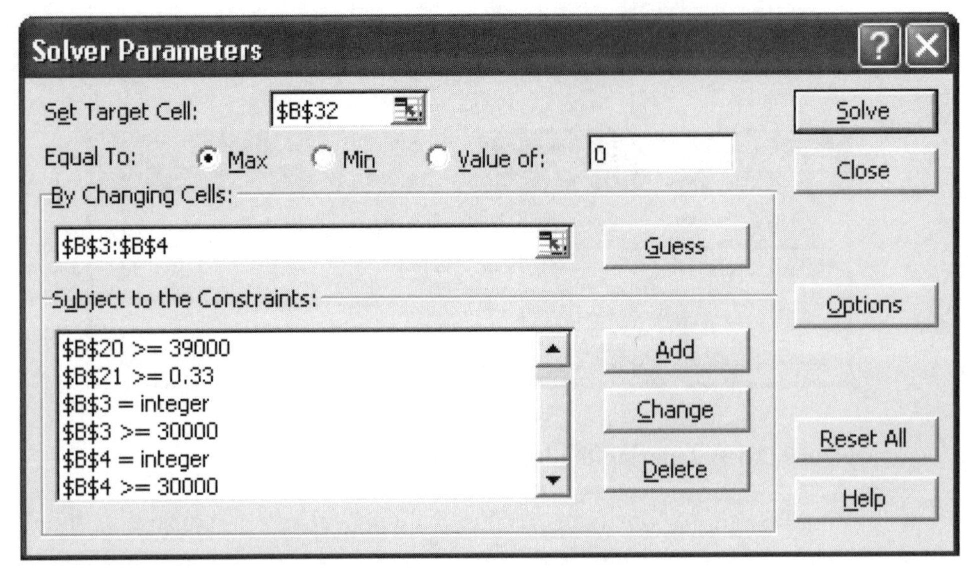

Figure D-21 Remainder of Extension Case constraints

The constraints are now only for the minimum production levels, the ratio of net income after taxes to total revenue, machine-hours shared resource constraints, and whole number output. When the Solver is run, the Answer Report looks like the one shown in Figure D-22.

Target Cell (Max)

Cell	Name	Original Value	Final Value
B32	NET INCOME AFTER TAXES	9.90	556198.38

Adjustable Cells

Cell	Name	Original Value	Final Value
B3	NUMBER OF BASKETBALLS	1	30000
B4	NUMBER OF FOOTBALLS	1	83333

Constraints

Cell	Name	Cell Value	Formula	Status
B20	TOTAL MACHINE HOURS USED (BB + FB)	39999.90	B20>=39000	Not Binding
B21	RATIO OF NET INCOME TO TOTAL REVENUE	0.416109655	B21>=0.33	Not Binding
B20	TOTAL MACHINE HOURS USED (BB + FB)	39999.90	B20<=40000	Not Binding
B3	NUMBER OF BASKETBALLS	30000	B3>=30000	Binding
B4	NUMBER OF FOOTBALLS	83333	B4>=30000	Not Binding
B3	NUMBER OF BASKETBALLS	30000	B3=integer	Binding
B4	NUMBER OF FOOTBALLS	83333	B4=integer	Binding

Figure D-22 Extension Case Answer Report

The Extension Case answer differs from the Base Case answer. Which production schedule should management use? The one that has maximum production limits? Or the one that has no such limits? These questions are posed to get you to think about the purpose of using a DSS program. Two scenarios, the Base Case and the Extension Case, were modeled in the Solver. The very different answers are shown in Figure D-23.

	Base Case	Extension Case
Basketballs	57,143	30,000
Footballs	38,095	83,333

Figure D-23 The Solver's answers for the two cases

Are you able to use this output alone to decide how many of each kind of ball to produce? No, you cannot. You must also refer to the "Target," which in this case is net income. Figure D-24 shows the answers with net income target data.

	Base Case	Extension Case
Basketballs	57,143	30,000
Footballs	38,095	83,333
Net Income	$473,143	$556,198

Figure D-24 The Solver's answers for the two cases—with target data

Viewed this way, the Extension Case production schedule looks better, because it gives you a higher target net income.

At this point, you should save the **SPORTS2.xls** file (File—Save) and then close it (File—Close).

Tutorial D

USING THE SOLVER ON A NEW PROBLEM

Here is a short problem that will let you test what you have learned about the Excel Solver.

Setting Up the Spreadsheet

Assume that you run a shirt-manufacturing company. You have two products: (1) polo-style T-shirts, and (2) dress shirts with button-down collars. You must decide how many T-shirts and how many button-down shirts to make. Assume that you'll sell every shirt you make.

AT THE KEYBOARD

Open a file called **SHIRTS.xls**. Set up a Solver spreadsheet to handle this problem.

CHANGING CELLS Section

Your changing cells should look like those shown in Figure D-25.

	A	B
1	**SHIRT MANUFACTURING EXAMPLE**	
2	**CHANGING CELLS**	
3	NUMBER OF T-SHIRTS	1
4	NUMBER OF BUTTON-DOWN SHIRTS	1

Figure D-25 Shirt manufacturing changing cells

CONSTANTS Section

Your spreadsheet should contain the constants shown in Figure D-26. A discussion of constant cells (and some of your company's operations) follows the figure.

	A	B
6	**CONSTANTS**	
7	TAX RATE	0.28
8	SELLING PRICE: T-SHIRT	8.00
9	SELLING PRICE: BUTTON-DOWN SHIRT	36.00
10	VARIABLE COST TO MAKE: T-SHIRT	2.50
11	VARIABLE COST TO MAKE: BUTTON-DOWN SHIRT	14.00
12	COTTON USAGE (LBS): T-SHIRT	1.50
13	COTTON USAGE (LBS): BUTTON-DOWN SHIRT	2.50
14	TOTAL COTTON AVAILABLE (LBS)	13000000
15	BUTTONS PER T-SHIRT	3.00
16	BUTTONS PER BUTTON-DOWN SHIRT	12.00
17	TOTAL BUTTONS AVAILABLE	110000000

Figure D-26 Shirt manufacturing constants

- The TAX RATE is .28 on pre-tax income, but no taxes are paid on losses.
- SELLING PRICE: You sell polo-style T-shirts for $8 and button-down shirts for $36.
- VARIABLE COST TO MAKE: It costs $2.50 to make a T-shirt and $14 to make a button-down shirt. These variable costs are for machine-operator labor, cloth, buttons, and so forth.
- COTTON USAGE: Each polo T-shirt uses 1.5 pounds of cotton fabric. Each button-down shirt uses 2.5 pounds of cotton fabric.
- TOTAL COTTON AVAILABLE: You have only 13 million pounds of cotton on hand to be used to make all the T-shirts and button-down shirts.
- BUTTONS: Each polo T-shirt has 3 buttons. By contrast, each button-down shirt has 1 button on each collar tip, 8 buttons down the front, and 1 button on each cuff, for a total of 12 buttons. You have 110 million buttons on hand to be used to make all your shirts.

CALCULATIONS Section

Your spreadsheet should contain the calculations shown in Figure D-27.

	A	B
19	**CALCULATIONS**	
20	RATIO OF NET INCOME TO TOTAL REVENUE	
21	COTTON USED: T-SHIRTS	
22	COTTON USED: BUTTON-DOWN SHIRTS	
23	COTTON USED: TOTAL	
24	BUTTONS USED: T-SHIRTS	
25	BUTTONS USED: BUTTON-DOWN SHIRTS	
26	BUTTONS USED: TOTAL	
27	RATIO OF BUTTON-DOWNS TO T-SHIRTS	

Figure D-27 Shirt manufacturing calculations

Calculations (and related business constraints) are discussed next.

- RATIO OF NET INCOME TO TOTAL REVENUE: The minimum return on sales (ratio of net income after taxes divided by total revenue) is .20.
- COTTON USED/BUTTONS USED: You have a limited amount of cotton and buttons. The usage of each resource must be calculated, then used in constraints.
- RATIO OF BUTTON-DOWNS TO T-SHIRTS: You think you must make at least 2 million T-shirts and at least 2 million button-down shirts. You want to be known as a balanced shirtmaker, so you think that the ratio of button-downs to T-shirts should be no greater than 4:1. (Thus, if 9 million button-down shirts and 2 million T-shirts were produced, the ratio would be too high.)

INCOME STATEMENT Section

Your spreadsheet should have the income statement skeleton shown in Figure D-28.

	A	B
29	**INCOME STATEMENT**	
30	T-SHIRT REVENUE	
31	BUTTON-DOWN SHIRT REVENUE	
32	TOTAL REVENUE	
33	VARIABLE COSTS: T-SHIRTS	
34	VARIABLE COSTS: BUTTON-DOWNS	
35	TOTAL COSTS	
36	INCOME BEFORE TAXES	
37	INCOME TAX EXPENSE	
38	NET INCOME AFTER TAXES	

Figure D-28 Shirt manufacturing income statement line items

The Solver's target is net income, which must be maximized.

Use the table shown in Figure D-29 to write out your constraints before entering them into the Solver.

Expression in English	Fill in the Excel Expression
Net income to revenue	_____ >= _____
Ratio of BDs to Ts	_____ <= _____
Min T-shirts	_____ >= _____
Min Button-Downs	_____ >= _____
Usage of buttons	_____ <= _____
Usage of cotton	_____ <= _____

Figure D-29 Logic of shirt manufacturing constraints

When you are finished with the program, print the sheets. Then, use File—Save, File—Close, and then File—Exit, to leave Excel.

Tutorial D

TROUBLE-SHOOTING THE SOLVER

Use this section to overcome problems with the Solver and as a review of some Windows file-handling procedures.

Rerunning a Solver Model

Assume that you have changed your spreadsheet in some way and want to rerun the Solver to get a new set of answers. (For example, you may have changed a constraint or a formula in your spreadsheet.) Before you click Solve again to rerun the Solver, you should put the number 1 in the changing cells. The Solver can sometimes give odd answers if its point of departure is a set of prior answers.

Creating Over-Constrained Models

It is possible to set up a model that has no logical solution. For example, in the second version of the sporting goods problem, suppose that you had specified that at least 1 million basketballs were needed. When you clicked Solve, the Solver would have tried to compute an answer, but then would have admitted defeat by telling you that no feasible solution is possible, as shown in Figure D-30.

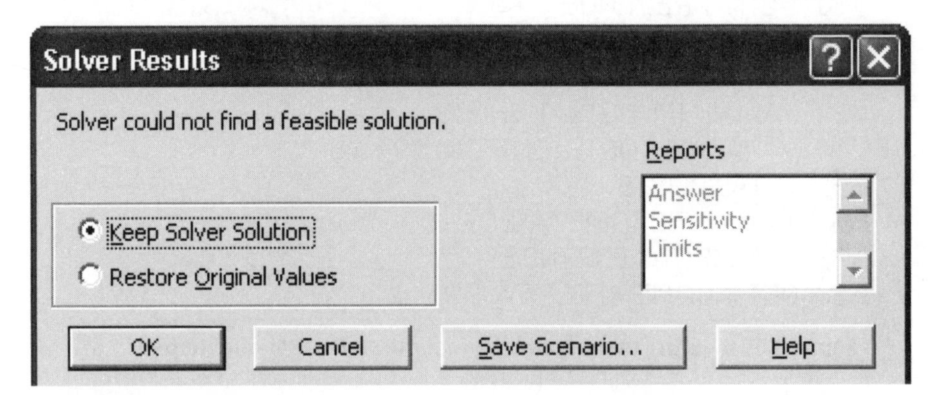

Figure D-30 Solver Results message: Solution not feasible

In the Reports window, the choices (Answer, etc.) would be in gray—indicating they are not available as options. Such a model is sometimes called "over-constrained."

Setting a Constraint to a Single Amount

It's possible you'll want an amount to be a specific number, as opposed to a number in a range. For example, if the number of basketballs needed to be exactly 30,000, then the "equals" operator would be selected, as shown in Figure D-31.

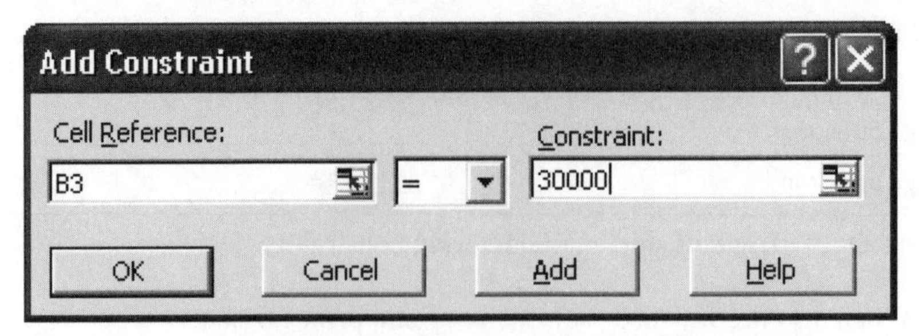

Figure D-31 Constraining a value to equal a specific amount

Setting a Changing Cell to an Integer

You may want to force changing cell values to be integers. The way to do that is to select the Int operator in the Add Constraint window. This was described in a prior section.

Forcing the Solver to find only integer solutions slows the Solver down. In some cases, the change in speed can be noticeable to the user. Doing this can also prevent the Solver from seeing a feasible solution—when one can be found if the Solver is allowed to find non-integer answers. For these reasons, it's usually best to not impose the integer constraint unless the logic of the problem demands it.

Deleting Extra Answer Sheets

Suppose that you've run different scenarios, each time asking for an Answer Report. As a result, you have a number of Answer Report sheets in your Excel file, but you don't want to keep them all. How do you get rid of an Answer Report sheet? Follow this procedure: First, get the unwanted Answer Report sheet on the screen by clicking the sheet's tab. Then select Edit—Delete Sheet. You will be asked if you really mean it. If you do, click accordingly.

Restarting the Solver with All-New Constraints

Suppose that you wanted to start over, with a new set of constraints. In the Solver Parameters window, click Reset All. You will be asked if you really mean it, as shown in Figure D-32.

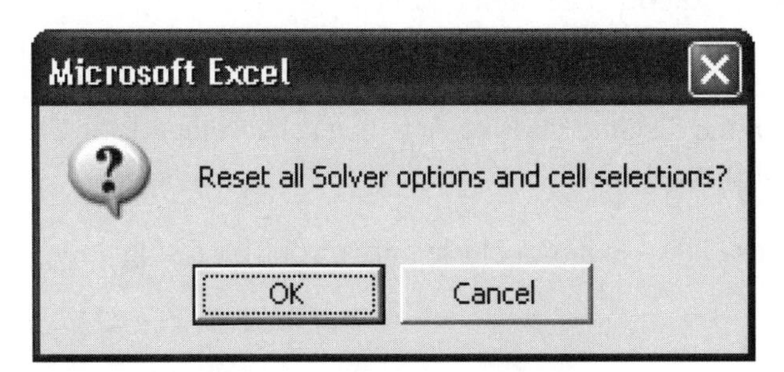

Figure D-32 Reset options warning query

If you do, then select OK. This gives you a clean slate, with all entries deleted, as shown in Figure D-33.

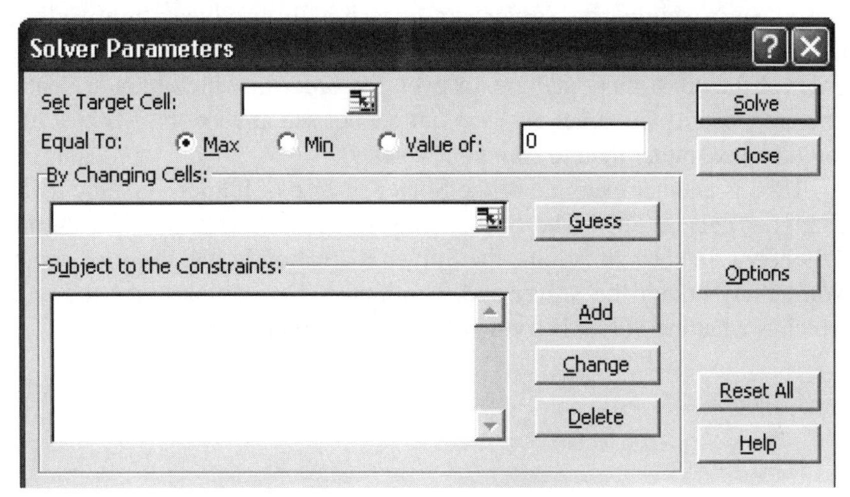

Figure D-33 Reset Solver Parameters window

As you can see, the target cell, changing cells, and constraints have been reset. From this point, you can specify a new model.

If you select Reset All, you really are starting over. If you merely want to add, delete, or edit a constraint, do not use Reset All. Use the Add, Delete, or Change buttons, as the case may be.

Printing Cell Formulas in Excel

To show the Cell Formulas on the screen, press the Ctrl and the left quote (`) keys at the same time: Ctrl-`. (The left quote is usually on the same key as the tilde [~].) This automatically widens cells so the formulas can be read. You can change cell widths by clicking and dragging at the column indicator (A, B, etc.) boundaries.

To print the formulas, just use File—Print. Print the sheet as you would normally. To restore the screen to its typical appearance (showing values, not formulas), press Ctrl-` again. (It's like a toggle switch.) If you did not change any column widths when in the cell formula view, the widths will be as they were.

Review of Printing, Saving, and Exiting Procedures

Print the Solver spreadsheets in the normal way. Activate the sheet, then select File—Print. You can print an Answer Report sheet in the same way.

To save a file, use File—Save, or File—Save As. Be sure to select drive A: in the Drive window, if you intend your file to be on a diskette. When exiting from Excel, always start with File—Close (with the diskette in drive A:), then select File—Exit. Only then should you take the diskette out of drive A:.

If you merely use File—Exit (not closing first), you risk losing your work.

Sometimes, you might think that the Solver has an odd sense of humor. For instance, your results might differ from the target answers that your instructor provides for a case. Thinking that you've done something wrong, you ask to compare your cell formulas and constraint expressions with those your instructor created. Lo and behold, you can see no differences! Surprisingly, the Solver can occasionally produce slightly different outputs from inputs that are seemingly the same, for no apparent reason. Perhaps, for your application, the order of the constraints matters, or even the order in which they are entered. In any case, if you are close to the target answers but cannot see any errors, it's best to see your instructor for guidance, rather than to spin your wheels.

Here is another example of the Solver's sense of humor. Assume that you ask for Integer changing cell outputs. The Solver may tell you that the correct output is 8.0000001, or 7.9999999. In such situations, the Solver is apparently just not sure about its own rounding! You merely humor the Solver and (continuing the example) take the result as the integer 8, which is what the Solver is trying to say in the first place.

The Golf Course Architecture Problem

8
CASE

Decision Support Using Excel Solver

Preview

A golf course architect is designing a new golf course, and he needs your advice. He wants to know how to configure the course to maximize golfers' enjoyment and to meet the town's needs. In this case, you will use the Excel Solver to provide the advice he needs.

Preparation

- Review spreadsheet concepts discussed in class and/or in your textbook.
- Complete any exercises that your instructor assigns.
- Complete any part of Tutorial D that your instructor assigns, or refer to it as necessary.
- Review file-saving procedures for Windows programs. These are discussed in Tutorial D.
- Refer to Tutorial E as necessary.

BACKGROUND

A wealthy golfer in your town recently died. He owned hundreds of acres on the outskirts of town, and in his will he gave the town 42 acres of the land on which to build an 18-hole public golf course. (A "public" course is one on which anyone can play who is willing to pay the greens fee.) The town can take its 42 acres out of any part of the land and in whatever configuration that the town desires. A small management team has been assembled in the town's Parks and Recreation Department. Their first official act has been to hire a golf course architect, Sandy Mulligan, to design the golf course.

In the design phase, a golf course architect decides how to configure the course. The architect must take into account the topography available (streams, trees, hills, and so forth) and the budget for earth-moving machines. In this case, there is no budget for artificial contouring, so Sandy Mulligan will just have to make do with what nature offers.

There are only 42 acres available for the golf holes themselves. Sandy Mulligan tells you that is a pretty tight space for a golf course these days. So, he has called you in to render some decision support.

Because you are not a golfer, Sandy Mulligan gives you a brief overview of golf and its peculiar vocabulary. For instance, a "hole" is a defined section of a golf course—an 18-hole golf course will have 18 of these sections. The beginning of a hole is a raised area called a "tee." Golfers initially strike their golf balls from this point. For each hole, a golfer's goal is to put a golf ball in a cup buried in an area of short manicured grass, called a "green," which is at the end of the hole. Swinging a golf club to strike the ball is called a "stroke." The object of the game is to use as few strokes as possible to move the ball from the tee to the cup in the green. "Par" is the number of strokes that a golfer should use to move the ball from the tee into the cup at the end of the hole.

Golf courses have holes that are rated as par three, par four, and par five. Par-three holes are usually 125 to 225 yards in length, from tee to green. Par-four holes are usually 300 to 450 yards in length. Par-five holes are usually 475 to 550 yards in length.

For a par-three hole, the golfer should strike the ball three times: once to get the ball in the air and onto the green, then twice with the special "putter" club to "putt" the golf ball into the cup. Similarly, in a par-four hole, the golfer should hit the ball four times: twice to get it onto the green and then two putts to get it into the cup. For a par-five hole, the golfer should hit the ball five times: three times to get it onto the green and then two putts to put it in the cup.

Par-three holes are always "straight" holes. That means that the golfer standing on the tee can see the green, which is only a relatively short distance away. Par threes vary by whether they are (roughly speaking) "long" or "short." For instance, a long par three might be more than 175 yards in length from tee to green, whereas a short par three might be fewer than 175 yards. Most golfers would say that a par-three hole that is 175 yards in length is a much harder hole to complete in par than a hole that is 140 yards in length.

By contrast, par-four and par-five holes are not always "straight" holes. Sometimes they have what is called a "dogleg" (apparently named after the shape of a dog's hind legs). The path to the hole bends to the right or to the left, about halfway to the green. Thus, a dogleg hole's green is not visible from the tee. Doglegs add variety, and difficulty, to the golfing experience, and most courses usually have at least one or two, but some courses have many.

The acreage that a hole requires is a function of its length, whether it has a dogleg (which tends to increase acreage needed), its width, the amount of forestry that lines the hole, and other factors. Sandy Mulligan tells you that he has calculated the number of acres that will be taken up by each kind of hole in the new course. His calculations are shown in Figure 8-1.

Acreage Required for Golf Holes	
Kind of hole	**Acreage taken up by hole**
Straight par 5	3.0
Dogleg par 5	3.5
Straight par 4	2.0
Dogleg par 4	2.5
Long par 3	1.0
Short par 3	0.75

Figure 8-1 Acreage required for golf holes

Sandy Mulligan says that surveys reveal that golfers who golf at public golf courses enjoy some kinds of holes more than others. The "enjoyability" index for each kind of hole is shown in Figure 8-2.

Golf Hole Enjoyability Index	
Kind of hole	**Enjoyability index**
Straight par 5	2.0
Dogleg par 5	1.5
Straight par 4	1.5
Dogleg par 4	2.0
Long par 3	1.75
Short par 3	2.25

Figure 8-2 Golf hole enjoyability index

Thus, most golfers enjoy a short par three the most, presumably because they can score closest to par on this kind of hole. Straight par fives are more enjoyable than dogleg par fives, presumably because dogleg par fives are so difficult to complete in par. But dogleg par fours, surprisingly, are more enjoyed than straight par fours.

Architects have some flexibility in how to set up a course. Of course, there must be a total of 18 holes in a full course. By tradition, courses have no more than four par-three holes and no more than four par-five holes. A full 18-hole course's par—the total of the pars for all 18 holes—can be 70, 71, or 72.

Sandy Mulligan has asked you to build a decision support model that will reveal how many of the six kinds of holes to include in the new course. *His goal is to maximize golfer enjoyability for the 18 holes.* Presumably, the more enjoyable a round of golf is at the new course, the more golfers will be playing there, and the more money the town will make.

Sandy Mulligan gives you some other rules to follow: First, the golf course (including clubhouse and parking lot) must use at least 36 acres. ("We are not building a putt-putt course here," Sandy Mulligan says.) Second, the course should have at least two straight par fours and at least two dogleg par fours. In addition, it should have at least one straight par five, one dogleg par five, one short par three, and one long par three. Thus, there can be no more than 14 par fours in the course. The clubhouse and the golf course parking lot will take up two of the total acres used.

The town's mayor has her own ideas to contribute. Instead of having a small clubhouse that just serves the needs of golfers, she thinks there might be some money to be made holding small conventions in a big clubhouse. The mayor claims, "Golfers would enjoy a big clubhouse, too. Furthermore, why should the big cities get all the conventions? Why shouldn't conventioneers come to *our* town to spend their money?" So the mayor proposes building a large, multipurpose clubhouse that could double as a convention center and accommodate conventioneers. This larger clubhouse and parking lot would take up four of the total acres available. This option might mean changes in the course layout, however—perhaps some of the longer holes would have to be dropped in favor of shorter holes to make the larger clubhouse fit.

ASSIGNMENT 1 CREATING A SPREADSHEET FOR DECISION SUPPORT

In this assignment, you will produce spreadsheets that model the business decision. In Assignment 1A, you will make a Solver spreadsheet to model the golf course design decision with a small clubhouse. This will be the Base Case. In Assignment 1B, the Extension Case, you will make a Solver spreadsheet to model the golf course design decision with a larger clubhouse.

In Assignments 2 and 3, you will use the spreadsheet models to develop information needed to recommend the best course design—including whether to build the larger clubhouse. In Assignment 2, you'll document your recommendations in a memorandum. In Assignment 3, you'll give your recommendations in an oral presentation.

Your spreadsheets for this assignment should have the sections that follow:

- **CHANGING CELLS**
- **CONSTANTS**
- **CALCULATIONS**

You will be shown how each section should be set up before entering cell formulas. Your spreadsheets will also include the decision constraints, which you'll enter using the Solver.

Assignment 1A: Creating the Spreadsheet—Base Case

A discussion of each spreadsheet section follows. The discussion is about (1) how each section should be set up and (2) the logic of the formulas in the section's cells. When you type in the spreadsheet skeleton, follow the order given in this section.

CHANGING CELLS Section

Your spreadsheet should have the changing cells shown in Figure 8-3.

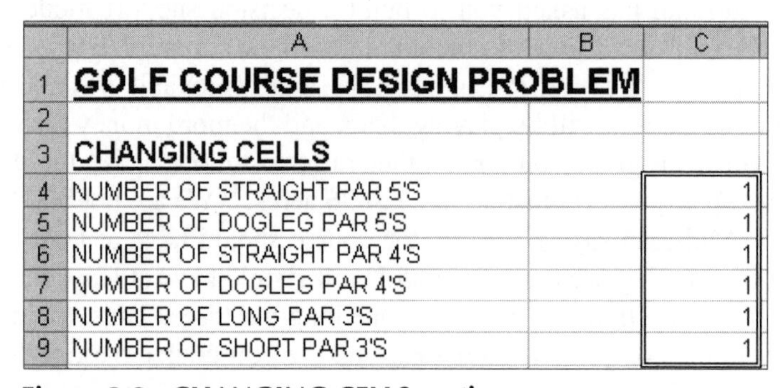

	A	B	C
1	GOLF COURSE DESIGN PROBLEM		
2			
3	CHANGING CELLS		
4	NUMBER OF STRAIGHT PAR 5'S		1
5	NUMBER OF DOGLEG PAR 5'S		1
6	NUMBER OF STRAIGHT PAR 4'S		1
7	NUMBER OF DOGLEG PAR 4'S		1
8	NUMBER OF LONG PAR 3'S		1
9	NUMBER OF SHORT PAR 3'S		1

Figure 8-3 CHANGING CELLS section

You are asking the Solver model to compute how many of each kind of hole to include in the course. Start with a "1" in each cell. The Solver will change each 1 as it computes the answer. Of course, you would not want the Solver to recommend a fractional part of a hole.

CONSTANTS Section

Your spreadsheet should have the constants shown in Figure 8-4. An explanation of the line items follows the figure.

	A	B	C
11	**CONSTANTS**		
12	ENJOYABILITY INDICES:		--
13	STRAIGHT PAR 5		2.00
14	DOGLEG PAR 5		1.50
15	STRAIGHT PAR 4		1.50
16	DOGLEG PAR 4		2.00
17	LONG PAR 3		1.75
18	SHORT PAR 3		2.25
19	ACREAGE PER HOLE:		--
20	STRAIGHT PAR 5		3.00
21	DOGLEG PAR 5		3.50
22	STRAIGHT PAR 4		2.00
23	DOGLEG PAR 4		2.50
24	LONG PAR 3		1.00
25	SHORT PAR 3		0.75
26	ACREAGE -- CLUBHOUSE		2.00

Figure 8-4 CONSTANTS section

- Each type of hole has a different enjoyability index.
- Each type of hole requires a different amount of acreage per hole.
- The clubhouse and parking lot are expected to require two acres.

Leave a few blank lines after the constants for Extension Case expansion.

CALCULATIONS Section

Your spreadsheet should calculate the amounts shown in Figure 8-5. Calculated values may be based on the values of the changing cells and/or the constants. An explanation of the line items follows Figure 8-5.

	A	B	C
28	**CALCULATIONS**		
29	ACREAGE USED:		--
30	STRAIGHT PAR 5		
31	DOGLEG PAR 5		
32	STRAIGHT PAR 4		
33	DOGLEG PAR 4		
34	LONG PAR 3		
35	SHORT PAR 3		
36	TOTAL ACREAGE USED		
37	TOTAL PAR		
38	TOTAL HOLES		
39	ENJOYABILITY:		--
40	STRAIGHT PAR 5		
41	DOGLEG PAR 5		
42	STRAIGHT PAR 4		
43	DOGLEG PAR 4		
44	LONG PAR 3		
45	SHORT PAR 3		
46	TOTAL ENJOYABILITY		
47	NUMBER OF PAR 4's		
48	NUMBER OF PAR 5's		
49	NUMBER OF PAR 3's		

Figure 8-5 CALCULATIONS section

- ACREAGE USED for each type of hole is a function of the number of that type used in the course and the acreage for that type of hole.
- TOTAL ACREAGE USED is the total for all holes and for the clubhouse and parking lot.
- TOTAL PAR is a function of the number of each type of hole included and that type's par per hole.
- The TOTAL HOLES for all types of holes should equal 18.
- ENJOYABILITY for each type of hole is a function of the number of that type of hole used in the course and the enjoyability index for that type of hole.
- TOTAL ENJOYABILITY is the sum for all holes included. Remember that the goal of the design is to maximize total enjoyability.
- The NUMBER OF PAR 4's depends on the number of straight and dogleg par fours included.
- The NUMBER OF PAR 5's depends on the number of straight and dogleg par fives included.
- The NUMBER OF PAR 3's depends on the number of short and long par threes included.

Constraints and Running the Solver

Determine the constraints. Enter the Base Case decision constraints, using the Solver. Run the Solver and ask for the Answer Report when the Solver says that a solution has been found that satisfies the constraints.

When you're done, print the entire workbook, including the Solver Answer Report sheet. Save the workbook. (File—Save; **GOLFBASE.xls** would be a good file name.) Then, to prepare for the Extension Case, use File—*Save As* to make a *new* spreadsheet. (**GOLFEXT.xls** would be a good file name.)

Assignment 1B: Creating the Spreadsheet—Extension Case

The mayor thinks that everyone's enjoyability would be enhanced if the clubhouse/convention center were built. It's not obvious that this is the case, but the mayor is the boss, and so her opinion is important. Sandy Mulligan has agreed to consider a new set of rules to see whether it's possible to allocate some land from the golf course to allow for the bigger clubhouse.

According to the mayor, the bigger clubhouse would add four "enjoyment" points over and above what playing the course would yield. Sandy Mulligan agrees that, if this way of thinking yields more total enjoyability per round than the Base Case scenario, he will build a course with the bigger clubhouse.

Adjust the spreadsheet to handle the new conditions. The clubhouse acreage constant will increase, and a constant for clubhouse enjoyability should be added. Run the Solver. Ask for the Answer Report when the Solver says that a solution has been found that satisfies the constraints. When you're done, print the entire workbook, including the Solver Answer Report sheet. Save the workbook when you're done. Close the file and exit Excel.

Assignment 2 Using the Spreadsheet for Decision Support

You have built the Base Case and Extension Case models because you want to know how many of each kind of golf hole to include in the course design and which size clubhouse to build. You will now complete the case by (1) using the Answer Report to gather the data you need to make the course design decisions and (2) documenting your recommendations in a memorandum.

Assignment 2A: Using the Spreadsheet to Gather Data

You have printed the Answer Report sheet for each scenario. Each sheet should tell how many of each kind of hole to include, the total par, and the target total enjoyability. Note that data for each scenario on a separate sheet of paper for inclusion in a table in your memorandum. The form of that table is shown in Figure 8-6.

Golf Course Feature	Base Case (Average Clubhouse)	Extension Case (Large Clubhouse)
Total Enjoyment Points		
Total Acreage		
Total Par		
Number of Straight Par 5's		
Number of Dogleg Par 5's		
Number of Straight Par 4's		
Number of Dogleg Par 4's		
Number of Long Par 3's		
Number of Short Par 3's		

Figure 8-6 Form for table in memorandum

Assignment 2B: Documenting Your Recommendations in a Memorandum

Write a brief memorandum to architect Sandy Mulligan about the results. Observe the following requirements:

- Your memorandum should have a proper heading (DATE/ TO/ FROM/ SUBJECT). You may wish to use a Word memo template (File—New, click Memos, choose Contemporary Memo, and click OK).
- Briefly outline the golf course construction decision. Then state which kind of clubhouse should be built. Total enjoyability should rule the decision. Refer the reader to the data table to see how many of each kind of hole should be included, total par, and expected total enjoyability.
- Include the summary of data table shown in Figure 8-6.

Enter your summary of data table into Word, using the following procedure:

1. Select the **Table** menu option, point to **Insert**, then click **Table**.
2. Enter the number of rows and columns.
3. Select **AutoFormat** and choose **Table Grid 1**.
4. Select **OK**, then select **OK** again.

Assignment 3 Giving an Oral Presentation

Sandy Mulligan was so impressed by your memorandum that he has asked you to give an oral presentation of your findings to the golf course's management team and the mayor. For your presentation, you'll want to include appropriate visual aids and handouts. Be prepared to defend your recommendations—the managers and the mayor want to be reassured that your plan makes the best use of the bequest. Your presentation should take approximately 10 minutes, including a question-and-answer period.

Deliverables

Assemble the following deliverables for your instructor:

1. Base Case and Extension Case spreadsheet printouts
2. Memorandum
3. Presentation visual aids as appropriate
4. Diskette, which should have your memorandum file and your Excel files

Staple your printouts together, with the memo on top. If your diskette holds more than just the files for this case, write a note to your instructor, stating the name of the **.xls** files.

The New Wave Mutual Fund Investment Mix Problem

9

CASE

Decision Support Using Excel Solver

Preview

Each month, the New Wave Mutual Fund receives money from investors. The Fund's money managers then try to select securities that will yield the maximum return on investors' money. In this case, you'll use the Excel Solver to give a fund manager advice about the best mix of securities to buy each month.

Preparation

- Review spreadsheet concepts discussed in class and/or in your textbook.
- Complete any exercises that your instructor assigns.
- Complete any part of Tutorial D that your instructor assigns, or refer to it as necessary.
- Review file-saving procedures for Windows programs. These are discussed in Tutorial D.
- Refer to Tutorial E as necessary.

BACKGROUND

The New Wave Mutual Fund is managed by Susan Smith, one of the new breed of aggressive money managers on Wall Street. Many people now look to Wall Street for ways to save for retirement, and they entrust their savings to people like Susan. As those people continue to invest in mutual funds, Susan expects to take in $10,000,000 per month to invest in securities. Senior Fund management says that Susan can keep some of that money in cash, but she must invest the bulk of it in the following four kinds of securities:

- U.S. government Treasury securities (called "Treasuries"), such as 30-year government bonds
- AAA-rated corporate bonds, such as 15-year bonds issued by the Exxon Corporation
- "Blue chip" stocks, such as shares of U.S. Steel common stock
- High-technology stocks, such as shares of Microsoft common stock

Rates of Return and Risk

Each kind of security has an expected "rate of return" and an expected level of risk. Assume that a rate of return is one of the following:

- The amount of cash received by the investor (in dividends or interest) divided by the cost of the investment, *or*
- The increase in the security's market price divided by the cost of the investment

Investors seeking higher rates of return must assume higher levels of risk, including the risk of losing principal. When the value of the security goes down after being purchased, this is called "a loss of principal." For example, suppose that an investor buys 1,000 shares of AOL stock at $80 per share, and then the price falls to $10 per share. The loss of principal is 1,000 multiplied by $70, or a loss of $70,000. Susan knows that an investor who faces a higher level of risk will demand a higher rate of return.

The Fund's Research Department has estimated the expected annual rate of return for the four kinds of securities and for cash. The rates are shown in Figure 9-1.

Type of Investment	Expected Annual Rate of Return
Treasuries	.03
AAA bonds	.04
Blue chip stocks	.08
High-tech stocks	.10
Cash	0

Figure 9-1 Expected annual rates of return, Base Case

For example, suppose that $1,000,000 is invested in Treasuries. During the year, 3% of that amount—$30,000—will be received on that investment. That $30,000 will be "revenue" in the Fund's income statement. Another example: Suppose that $1,000,000 is invested in blue chip stocks. The prices of the stocks go up an average of 8% per year, yielding $80,000 of revenue in the Fund's income statement. Notice that the rates of return cited here are *annual* rates.

The Fund's Research Department has advised Susan of the expected levels of risk for different kinds of securities. Risk is denoted by an Expected Risk Indicator, which is an integer. The higher the indicator number, the higher the expected risk. The risk indicators are shown in Figure 9-2.

Type of Investment	Expected Risk Indicator
Treasuries	0
AAA bonds	1
Blue chip stocks	2
High-tech stocks	6
Cash	0

Figure 9-2 Expected Risk Indicators

As shown in Figure 9-2, Treasuries have zero risk because the U.S. government can be expected to repay its debts. AAA bonds have some risk, but not much, since the best companies almost always pay their debts. Blue chip stocks have some risk—the risk on stock prices is generally higher than on bond prices. High-tech stock prices are quite volatile, and their prices often go down. Sometimes they go down very much, very fast. In fact, sometimes these companies become insolvent, and the stock price goes to zero, resulting in a complete loss of principal. The Research Department has assigned zero risk to cash.

Expenses

Operating the Fund entails incurring expenses. First, the Fund must work with a separate management company that keeps computerized records, provides frequent reports from Wall Street, and so on. New Wave's management company charges New Wave a flat 1% per year multiplied by amounts invested in non-cash securities.

In addition, each kind of security must be watched by Susan and her analysts. The cost of doing this (analysts' salaries, office expenses, and so on) depends on the kind of security. Because high-risk securities require more oversight, expenses associated with high-risk securities are higher than for low-risk securities. The Fund's Research Department has estimated those expense factors for securities and cash. They are shown in Figure 9-3.

Type of Investment	Expense Factor
Treasuries	.001
AAA bonds	.003
Blue chip stocks	.004
High-tech stocks	.008
Cash	0

Figure 9-3 Investment expense factors

For example, suppose that $1,000,000 is invested in Treasuries. During the year, one-tenth of a percent of that sum—$1,000—will be spent monitoring the million dollars. Thus, that $1,000 will be a "variable" expense in New Wave's income statement for the year.

Susan has been given certain constraints by the Fund's senior management: The "net income to total investment" ratio must be at least 4%. That is, the Fund's net income after taxes, divided by the amount of dollars invested, must be at least .04. For example, if yearly net income after taxes is $1,000,000 on $10,000,000 invested, then the net income to total investment would be 10%, and management would be very satisfied. By contrast, if yearly net income after taxes is $300,000 on the $10,000,000 invested, Susan's management would not be satisfied with the 3% ratio.

Investment Requirements

Each month, Susan will get $10,000,000 to invest. She can invest no more than that each month, of course. Senior Fund management says that each month Susan must "park" at least $50,000 of that amount in a cash bank account (which earns no interest but has no risk) as a buffer against unexpected expenses. However, she can park no more than $1,000,000 per month; thus, she must invest at least $9,000,000 per month in securities.

Susan has been told that each month she must buy at least $1,000,000 of each kind of the four non-cash securities. Thus, she cannot buy all Treasuries to guarantee a return without risk, nor can she buy all high-tech stocks to guarantee high rates of return. She must diversify.

As a brake on Susan's known aggressiveness, Fund management has also told Susan that at least half of the total invested (including cash) must be in a combination of Treasuries, AAA bonds, and blue chip stocks.

Management has also set quantifiable risk-level boundaries for Susan. Fund management wants her to take some risk, but not too much risk. The "weighted average risk level" on the total invested each month (including cash) must be at least 1.5, but it should not exceed 3.5. The weighted average risk level is computed by the following procedure:

- weighting the risk indicators by the amount invested in each kind of security,
- adding the weighted amounts,
- then dividing that total by the total amount invested.

For illustrative purposes, here is a simplified example: Suppose that only Treasuries and high-tech stocks are purchased. Suppose that $5,000,000 is invested in Treasuries and $5,000,000 in high-tech stocks. The weighted average risk level would be 3.0:

(($5,000,000 * 0) + ($5,000,000 * 6)) / $10,000,000 = 3.0

(*Note*: In this simplified example, only two kinds of securities are considered, but Susan will actually have to invest in more than two kinds of securities.)

Susan thinks that the scenario outlined here is a reasonable one for the foreseeable future. However, some of New Wave's management think that even harder times are coming to Wall Street, and they are calling for even more conservatism. These managers want to consider different rules for Susan, which will be presented later, in the Extension Case.

Susan needs a plan for her investments that will maximize the Fund's net income without too much risk. She turns to you for decision support because she knows that you can model this problem in the Solver.

ASSIGNMENT 1 CREATING A SPREADSHEET FOR DECISION SUPPORT

In this assignment, you will produce spreadsheets that model the business decision. In Assignment 1A, you will make a Solver spreadsheet to model the mutual fund investment decision. This will be the Base Case. In Assignment 1B, the Extension Case, you will make a Solver spreadsheet to model the investment decision, given more conservative operating rules.

In Assignments 2 and 3, you will use the spreadsheet models to develop information needed to recommend the best investment mix for the mutual fund. In Assignment 2, you'll document your recommendations in a memorandum; in Assignment 3, you'll give your recommendations in an oral presentation.

Your spreadsheets for this assignment should have the sections that follow:

- **CHANGING CELLS**
- **CONSTANTS**
- **CALCULATIONS**
- **INCOME STATEMENT**

You will be shown how each section should be set up before entering cell formulas. Your spreadsheets will also include the decision constraints, which you'll enter using the Solver.

Assignment 1A: Creating the Spreadsheet—Base Case

A discussion of each spreadsheet section follows. The discussion is about (1) how each section should be set up and (2) the logic of the formulas in the section's cells. When you type in the spreadsheet skeleton, follow the order given in this section.

CHANGING CELLS Section

Your spreadsheet should have the changing cells shown in Figure 9-4.

	A	B	C
1	NEW WAVE MUTUAL FUND INVESTMENT MIX PROBLEM		
2			BASE CASE
3	CHANGING CELLS		
4	DOLLARS IN TREASURIES		1
5	DOLLARS IN AAA BONDS		1
6	DOLLARS IN BLUE CHIPS		1
7	DOLLARS IN HIGH-TECHS		1
8	DOLLARS IN CASH		1

Figure 9-4 CHANGING CELLS section

You are asking the Solver model to compute how many dollars to invest in each kind of security each month. Start with a "1" in each cell. The Solver will change each 1 as it computes the answer. It will be acceptable to Susan if the Solver recommends a fractional part of a dollar for a security.

CONSTANTS Section

Your spreadsheet should have the constants shown in Figure 9-5. An explanation of the line items follows the figure.

	A	B	C
10	**CONSTANTS**		
11	TAX RATE		0.280
12	RATE OF RETURN:		---
13	TREASURIES		0.030
14	AAA BONDS		0.040
15	BLUE CHIPS		0.080
16	HIGH-TECHS		0.100
17	CASH		0.000
18	MANAGEMENT FEE %		0.010
19	RISK FACTOR ASSIGNED:		---
20	TREASURIES		0.000
21	AAA BONDS		1.000
22	BLUE CHIPS		2.000
23	HIGH-TECHS		6.000
24	CASH		0.000
25	EXPENSE FACTOR:		---
26	TREASURIES		0.001
27	AAA BONDS		0.003
28	BLUE CHIPS		0.004
29	HIGH-TECHS		0.008
30	CASH		0.000

Figure 9-5 CONSTANTS section

- The TAX RATE on pre-tax income is 28%. Taxes are not paid on pre-tax losses, however.
- The RATE OF RETURN is different for each kind of security.
- The MANAGEMENT FEE of 1% is a flat percentage of the amount invested, except that the fee is not charged on cash put in a bank. (There is nothing to "manage" in that case.)
- The RISK FACTOR ASSIGNED refers to each kind of security's risk indicator.
- The EXPENSE FACTOR varies according to the kind of security. Dollar expenses in the income statement would be the factor multiplied by the amount invested in that type of security.

CALCULATIONS Section

Your spreadsheet should calculate the amounts shown in Figure 9-6. They will be used in the **INCOME STATEMENT** section and/or the **CONSTRAINTS** section. Calculated values may be based on the values of the changing cells and/or the constants and/or other calculations. An explanation of the line items follows Figure 9-6.

	A	B	C
33	**CALCULATIONS**		
34	REVENUE:		---
35	TREASURIES		
36	AAA BONDS		
37	BLUE CHIPS		
38	HIGH-TECHS		
39	CASH		
40	VARIABLE EXPENSES:		---
41	TREASURIES		
42	AAA BONDS		
43	BLUE CHIPS		
44	HIGH-TECHS		
45	CASH		
46	WEIGHTED AVG RISK LEVEL		
47	TOTAL DOLLARS INVESTED IN NON-CASH SECURITIES		
48	TOTAL INVESTED		
49	TOTAL DOLLARS INVESTED IN TREASURIES, AAA, BLUE CHIPS		
50	NET INCOME TO TOTAL INVESTED RATIO		
51	PERCENT IN TREASURIES, AAA, BLUE CHIPS		

Figure 9-6 CALCULATIONS section

- Compute the REVENUE for each security. Revenue is a function of the amount invested in a month and the security's annual rate of return. For a security, compute how much revenue will be earned in the year on the total invested in the month.
- Compute each security's VARIABLE EXPENSES, which are a function of the amount invested during a month and the security's expense ratio.
- Compute the WEIGHTED AVG RISK LEVEL for the five investments (including cash), following the logic of the example previously given.
- Compute the TOTAL DOLLARS INVESTED IN NON-CASH SECURITIES in a month.
- Compute the TOTAL INVESTED, including cash.
- Compute the TOTAL DOLLARS INVESTED IN TREASURIES, AAA, and BLUE CHIPS. Recall that there is a minimum that must be invested in these kinds of securities.
- Compute the NET INCOME TO TOTAL INVESTED RATIO. Recall that this ratio must be at least a certain amount for management to be happy with Susan's work.

INCOME STATEMENT Section

The statement shown in Figure 9-7 is the projected net income for a year on the amount invested in one month. An explanation of the line items follows Figure 9-7.

	A	B	C
54	**INCOME STATEMENT**		
55	TOTAL REVENUE		
56	EXPENSES:		---
57	MANAGEMENT FEE		
58	VARIABLE EXPENSES		
59	TOTAL EXPENSES		
60	INCOME BEFORE TAXES		
61	INCOME TAXES		
62	NET INCOME AFTER TAXES		

Figure 9-7 INCOME STATEMENT section

- The TOTAL REVENUE is obtained by totaling the revenues previously calculated.
- The MANAGEMENT FEE is a function of the management fee expense as applied to the non-cash amount invested.
- VARIABLE EXPENSES are obtained by totaling the variable expenses previously calculated.
- INCOME BEFORE TAXES is total revenue less total expenses.
- INCOME TAXES are zero if income before taxes is zero or less; otherwise, apply the tax rate multiplied by income before taxes.
- NET INCOME AFTER TAXES is income before taxes less income taxes.

Constraints and Running the Solver

Determine the constraints. Enter the Base Case decision constraints, using the Solver. Susan will not object if the Solver tells her to invest $1,500,000.***50*** in a type of security. Susan wants the plan to maximize net income after taxes, subject to the various investment and risk constraints. Run the Solver and ask for the Answer Report when the Solver says that a solution has been found that satisfies the constraints.

When you're done, print the entire workbook, including the Solver Answer Report sheet. Save the workbook. (File—Save; **FUNDBASE.xls** would be a good file name.) Then, to prepare for the Extension Case, use File—*Save As* to make a *new* spreadsheet. (**FUNDEXT.xls** would be a good file name.)

Assignment 1B: Creating the Spreadsheet—Extension Case

Some of the Fund's senior management and the Fund's researchers think that much harder times are coming to Wall Street. They think that the Fund needs more conservative investing policies. Susan has been told to think about a more conservative investment scenario. The possible changes are discussed next.

Susan must park at least $100,000 per month in cash, not $50,000, for contingencies. Susan has been told to consider lower rates of return, as shown in Figure 9-8.

Type of Investment	Expected Rate of Return
Treasuries	.025
AAA bonds	.035
Blue chip stocks	.07
High-tech stocks	.08
Cash	0

Figure 9-8 Expected annual rates of return, Extension Case

In addition, the weighted average risk level on the money invested each month would be at least 2.0, but would not exceed 3.0. Note that these limits are narrower and more conservative than those previously given.

Susan would be required to put at least 55% (not 50%) of the money in Treasuries, AAA bonds, and blue chip stocks combined. Recognizing that money is not as easy to make on Wall Street as it used to be, her required net income to total investment ratio would be at least 3.5% (versus 4% in the Base Case).

In addition, Fund management knows of another management company that would charge a flat rate of $75,000 for the year. This would probably be less costly than the Base Case arrangement for the year.

Modify the Extension Case spreadsheet to handle the more conservative scenario. Run the Solver. Ask for the Answer Report when the Solver says that a solution has been found that satisfies the constraints. When you're done, print the entire workbook, including the Solver Answer Report sheet. Save the workbook when you're done. Close the file and exit Excel.

Assignment 2 Using the Spreadsheet for Decision Support

You have built the Base Case and Extension Case models because you want to know the investment mix for each scenario and which scenario yields the highest net income after taxes, consistent with perceived risks. You will now complete the case by (1) using the Answer Reports to gather the data you need to make the investment mix decisions and (2) documenting your recommendations in a memorandum.

Assignment 2A: Using the Spreadsheet to Gather Data

You have printed the Answer Report sheet for each scenario. Each sheet tells how many dollars of each kind of security to purchase in a month, plus the target net income and risk in each case. Note that data for each scenario on a separate sheet of paper for inclusion in a table in your memorandum. The form of that table is shown in Figure 9-9.

	Base Case	Extension Case	*Difference*
Net Income After Taxes (in dollars)			
Net Income to Total Investment Ratio			
Weighted Average Risk Ratio			

Figure 9-9 Form for table in memorandum

Assignment 2B: Documenting Your Recommendations in a Memorandum

Write a brief memorandum to Susan about the results. Observe the following requirements:

- Your memorandum should have a proper heading (DATE/ TO/ FROM/ SUBJECT). You may wish to use a Word memo template (File—New, click Memos, choose Contemporary Memo, and click OK).
- Briefly outline the investment mix decision and then recommend the policy to adopt. To make your recommendation, compare net income after taxes in each case. Assume that the more conservative Extension Case policy would probably be adopted if the net income to total investment ratio is within 5% of the Base Case's ratio, and if the risk ratio is less in the Extension Case. Refer Susan to the data table to see the results.
- Include the summary of data table shown in Figure 9-9.

Enter your summary of data table into Word, using the following procedure:

1. Select the **Table** menu option, point to **Insert**, then click **Table**.
2. Enter the number of rows and columns.
3. Select **AutoFormat** and choose **Table Grid 1**.
4. Select **OK**, and then select **OK** again.

Assignment 3 Giving an Oral Presentation

Assume that Susan was so impressed by your memorandum that she has asked you to give an oral presentation of your findings to the Fund's senior management. For your presentation, you'll want to include appropriate visual aids. Be prepared to defend your recommendations—the senior Fund managers want to be reassured about your findings. Fund managers may want to "help" Susan decide which policy to adopt, given what is at that moment expected for the economy and the financial markets. Your presentation should take approximately 10 minutes, including a question-and-answer period.

Deliverables

Assemble the following deliverables for your instructor:

1. Base Case and Extension Case spreadsheet printouts
2. Memorandum
3. Presentation visual aids and handouts as appropriate
4. Diskette, which should have your memorandum file and your Excel files

Staple your printouts together, with the memo on top. If your diskette holds more than just the files for this case, write a note to your instructor, stating the names of the **.xls** files.

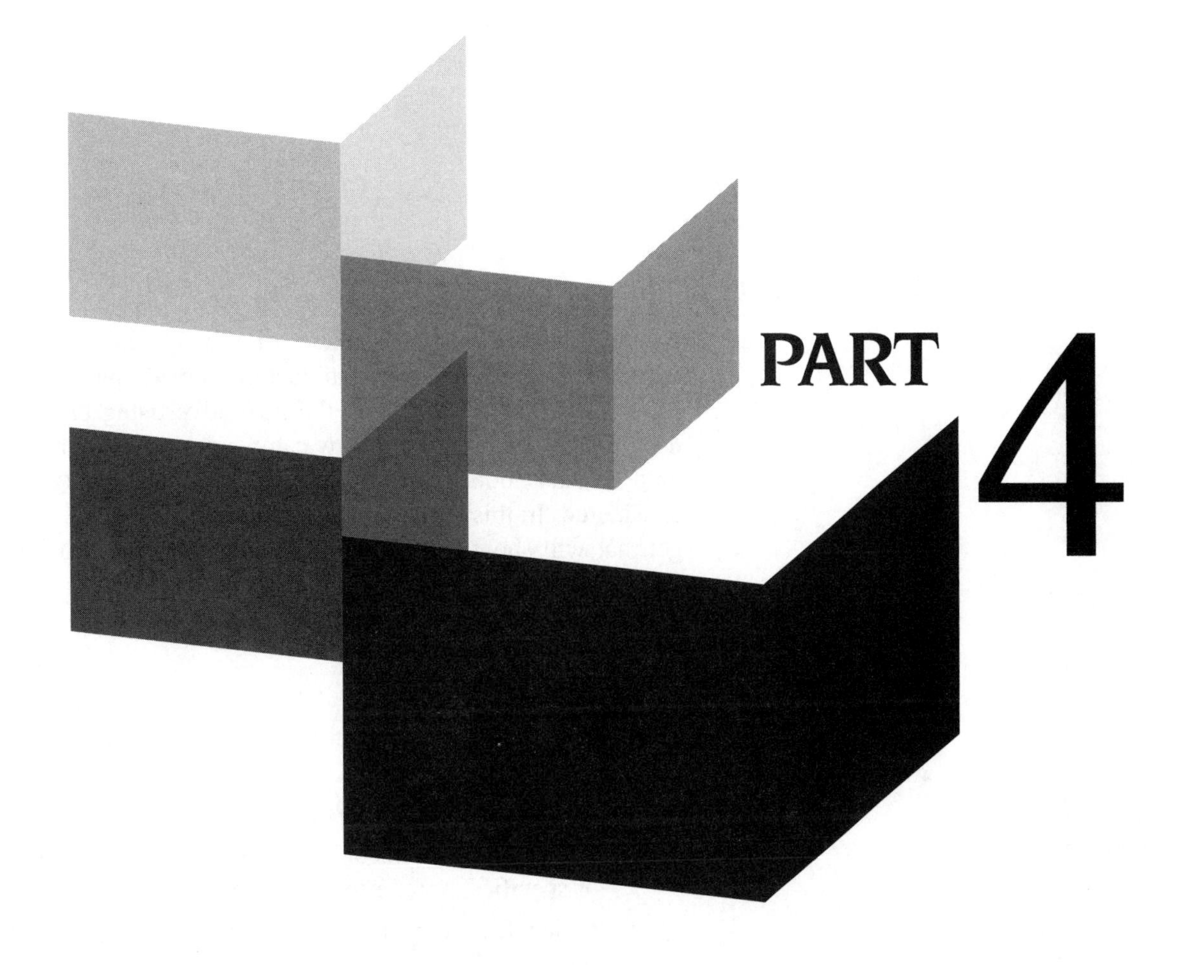

Decision Support Cases Using Basic Excel Functionality

The BigBiz Cash Budget Decision

DECISION SUPPORT USING EXCEL'S "WHAT IF?" CAPABILITY

PREVIEW

A manufacturer wants to launch a radically different six-month advertising campaign. The manufacturer already has commitments for its 2005 advertising, and the new advertising campaign will take time to develop, so the new campaign would not be rolled out until January 2006. Two ways of financing the campaign are being considered. In this case, you will use Excel to prepare a cash budget that will show the effects of the advertising campaign and its funding.

PREPARATION

- Review spreadsheet concepts discussed in class and/or in your textbook.
- Complete any exercises that your instructor assigns.
- Review any portions of Tutorials C and D that your instructor specifies, or refer to them as necessary.
- Review file-saving procedures for Windows programs. These are discussed in Tutorial D.
- Refer to Tutorial E as necessary.

➤ BACKGROUND

BigBiz company makes and sells industrial-grade freens, which are components in many industrial products. There are many freen makers, so competition is fierce. In any business, competition can result in lower selling prices and lower unit sales. To strengthen brand awareness, BigBiz management plans an intense and radically different advertising campaign in January 2006 through June 2006. Management thinks that the advertising campaign will increase unit sales somewhat and will also allow BigBiz to increase the selling price of freens.

Management is considering two ways of paying for the advertising campaign. Before undertaking the advertising campaign, management wants to explore the cash flow implications of each way. You have been asked to prepare a month-by-month 2006 cash budget for BigBiz.

Ways of Improving Cash Flow

The advertising campaign will require a substantial cash outlay in the first half of the year 2006. The company's treasurer says that two changes in cash flow could help pay for the campaign. One change would be to conserve cash by delaying payment of accounts payable. The second change would be to accelerate collection of accounts receivable by selling them to a "factoring" company. These alternatives are discussed next.

Delaying Payment of Accounts Payable

Currently, BigBiz purchases raw materials for the following month's production. When an order is placed, BigBiz records the payable owed to the vendor and then pays the accounts payable in the next month, a one-month time lag. Thus, materials for March 2006 production would be ordered in February 2006 and paid for in March 2006.

The treasurer points out that the company could adopt a new policy for accounts payable, effective January 2006. The policy would be to institute a two-month payment lag, or even a three-month lag, for accounts payable. Thus, materials for March 2006 production would still be ordered in February, but with a two-month payment lag the debt would be paid in April 2006. With a three-month lag, the debt would be paid in May 2006. The treasurer says that with a two-month lag, BigBiz would, in effect, be "borrowing" an amount equal to a month's payables from suppliers. With a three-month lag, the "borrowed" amount would equal two month's payables.

Vendor reaction is difficult to predict. It would take a while for vendors to realize that BigBiz was consistently delaying payments. Once that was realized, vendors might raise their selling prices. On the other hand, they might be persuaded that BigBiz would return to a one-month lag soon and not raise their prices.

This plan does have one predictable downside, however. At December 31, 2005, BigBiz is expected to owe its bank $20 million. The treasurer points out that the debt covenant specifies an interest rate that is tied to the company's accounts payable: The higher that the payables are, the higher the interest rate will be. Thus, increasing accounts payable would improve the cash position in one way but weaken it with respect to interest expense payments.

Selling Accounts Receivable to a Factoring Company

All of BigBiz's freen sales are on account—i.e., there are no cash sales. When BigBiz makes a sale, an accounts receivable is recorded. BigBiz collects receivables on this schedule:

- 10% in the month of sale
- 75% in the month after the sale
- 12% two months after the sale

After two months, BigBiz collects nothing, so 3% of sales go uncollected. Example: Suppose that BigBiz has $1 million sales in January 2006. The sum of $100,000 will be collected in January, $750,000 will be collected in February, and $120,000 will be collected in

March—$30,000 would go uncollected. (For a variety of reasons, it's typical in industry for some account receivables to go uncollected. For example, some customers may be unable to pay, and the seller must simply write off the account receivable as a bad debt. Sometimes, there are disputes about product quality, and the customer refuses to pay full price. Sometimes, items are returned—the customer doesn't pay for the item and it can't be resold.)

A "factor" is a company that buys another company's accounts receivable. The factor pays the seller for the receivables in the month of sale, so the selling company does not have to wait for its money or work at collecting accounts receivable. The factor does not pay the seller full face value, however. A factor might pay 80% of the receivables' face value. The factor tries to collect all the money originally due the seller. The factor makes a profit by collecting more from accounts receivable than was paid to the seller.

The treasurer says that a factoring company has offered to buy BigBiz's accounts receivable for 85% of face value, starting in January 2006 and going through December 2006. The trade-off, the treasurer says, is this: BigBiz collects less of its receivables in the long run, but in the short run much more of the receivables are collected in the month of sale. Another benefit is that BigBiz no longer must track down slow-paying customers. The factoring company would own the receivables and would be responsible for doing that.

The treasurer has called you in to create a month-by-month cash budget for 2006, assuming that the advertising campaign is undertaken under the two financing alternatives (factoring and delaying payables). The company's banker will finance any cash shortfalls in the period, and the treasurer wants to know how much BigBiz might need to borrow. Maximizing cash on hand and minimizing borrowings are the treasurer's goals.

ASSIGNMENT 1 CREATING A SPREADSHEET FOR DECISION SUPPORT

In this assignment, you will produce a spreadsheet that models the advertising campaign financing decision. In Assignments 2 and 3, you will use the spreadsheet model to develop information needed to decide how to finance the advertising campaign. In Assignment 2, you'll document your findings in a memorandum; in Assignment 3, you'll make an oral presentation.

Your spreadsheet for this assignment should have the sections that follow. You will be shown how each section should be set up before entering cell formulas.

- **CONSTANTS**
- **INPUTS**
- **SUMMARY OF KEY RESULTS**
- **CALCULATIONS**
- **CASH BUDGET**
- **DEBT OWED**

Assignment 1A: Creating the Spreadsheet

A discussion of each spreadsheet section follows. The discussion is about (1) how each section should be set up and (2) the logic of the formulas in the section's cells. When you type in the spreadsheet skeleton, follow the order given in this section.

CONSTANTS Section

Your spreadsheet should have the constants shown in Figure 10-1. Because cash collections in a month depend on sales in prior months, data for the last three months of 2005 are shown, along with data through March 2006. Remaining data are discussed in the comments following Figure 10-1.

	A	B	C	D	E	F	G
1	BIGBIZ CASH BUDGET						
2							
3	CONSTANTS	OCT	NOV	DEC	JAN	FEB	MAR
4	SELLING PRICE / UNIT	12	12	12	12	12	12
5	NUMBER OF UNITS SOLD	97000	99000	100000	100000	101000	102000
6	ADMINISTRATIVE EXPENSES	NA	NA	NA	20000	20000	20000
7	ADVERTISING EXPENSE	NA	NA	NA	800000	400000	100000
8	MINIMUM CASH DESIRED	NA	NA	NA	10000	10000	10000

Figure 10-1 CONSTANTS section

- SELLING PRICE / UNIT: BigBiz sells a freen for $12 per unit. The price will persist through June 2006. The company thinks that the price can be raised to $13 per unit in July 2006, after the advertising campaign is over, and that the price would stay at $13 for the rest of the year.
- NUMBER OF UNITS SOLD: Expected sales in the last three months of 2005 are shown; 100,000 freens are expected to be sold in January 2006. Sales are expected to rise 1,000 units each month. Expected sales in December 2006 are 111,000 units, rising to 112,000 units in January 2007. (*Note*: Assume that freens sold in a month were produced in that month.)
- ADMINISTRATIVE EXPENSES: These are incurred no matter what the production level and will total $20,000 per month in 2006. This includes expected income tax payments.
- ADVERTISING EXPENSE: The advertising campaign is expected to cost a lot early in the campaign, and then the amounts paid will decrease rapidly through June. Cash paid for advertising will be $800,000 in January, $400,000 in February, $100,000 in March, $25,000 in April, $15,000 in May, and $10,000 in June.
- MINIMUM CASH DESIRED: BigBiz needs $10,000 in cash in the bank to start any month. The company's banker will lend it enough to get to $10,000 if there is a cash shortfall at the end of any month.

INPUTS Section

Your spreadsheet should have the inputs shown in Figure 10-2. Each input would apply to all months in the campaign. The spreadsheet user is allowed to change an input to play "what if" with the analysis.

10	INPUTS		ALL MOS.
11	FACTOR (Y = YES, N = NO)	NA	
12	PAYMENT LAG (1, 2, 3 MONTHS)	NA	

Figure 10-2 INPUTS section

- FACTOR: If the factor is to be used, the user enters **Y** for "Yes"; otherwise, the user enters **N** for "No."
- PAYMENT LAG: If the accounts payable lag is to remain at one month, the user enters the digit **1**. If the lag is to be extended to two months, the user enters the digit **2**. If the lag is to be extended to three months, the user enters the digit **3**.

SUMMARY OF KEY RESULTS Section

Figure 10-3 shows the key results, which are merely echoed from elsewhere in the spreadsheet.

	A	B	C	D
14	SUMMARY OF KEY RESULTS		JUN	DEC
15	CASH ON HAND	NA		
16	DEBT OWED	NA		

Figure 10-3 SUMMARY OF KEY RESULTS section

Show the CASH ON HAND at the end of June and December 2006. Show DEBT OWED to the bank at the end of June and December 2006. Of course, the treasurer hopes that the financial position has improved in December as a result of the advertising campaign!

CALCULATIONS Section

Figure 10-4 shows amounts that should be calculated. Calculations are used in other calculations and/or in the cash budget. Use absolute addressing properly. Calculated values may be based on the values of the constants and/or other calculations. Calculations through March 2006 only are shown, but you should compute these amounts through December 2006. An explanation of the line items follows the figure.

	A	B	C	D	E	F	G
18	CALCULATIONS	OCT	NOV	DEC	JAN	FEB	MAR
19	SALES IN MONTH						
20	COLLECTIONS IN MONTH:	--	--	--	--	--	--
21	FROM FACTOR (85%)	NA	NA	NA			
22	FROM TWO MONTHS PRIOR (12%)	NA	NA	NA			
23	FROM MONTH PRIOR (75%)	NA	NA	NA			
24	FROM THIS MONTH (10%)	NA	NA	NA			
25	TOTAL COLLECTIONS	NA	NA	NA			
26	PURCHASES OF RAW MATERIALS						
27	ACCOUNTS PAYABLE						
28	PAYMENTS FOR PURCHASES	NA					
29	INTEREST RATE	NA	NA	NA			

Figure 10-4 CALCULATIONS section

- SALES IN MONTH: Compute dollar sales in each month. Sales are a function of selling price and units sold, both constants. You need to compute sales for October through December 2005 for use in computing cash collections.
- COLLECTIONS IN MONTH: Compute cash collections in a month. If the factor is used, collections are 85% of the month's sales. If the factor is *not* used, sales are collected over a three-month period. (*Note*: If the factor is used, January and February 2006 collections would still benefit from sales made in 2005.)
- PURCHASES OF RAW MATERIALS: Purchases in a month are based on the next month's unit sales. Purchases of raw materials cost $7 per unit; i.e., if 100,000 freens are to be made in May, the purchases would be in April for $700,000. (*Note*: You need to compute purchases for October through December 2005.)

- ACCOUNTS PAYABLE: These are a function of the payment lag input. If the lag remains one month, payables in a month equal the purchases for that month. (Example: Purchase materials in April for May production; pay for purchases in May.) However, if the lag exceeds one month, payables are put off. (*Note*: You can compute payables using an =If() function, but your instructor may instruct you to use the =Choose() function in some of the cells.)
- PAYMENTS FOR PURCHASES: Similarly, payments made are a function of the payment lag input. If the lag remains one month, payables in a month equal the purchases for the prior month. (Example: Purchase goods in April for May production; pay for those goods in May.) However, if the lag exceeds one month, payments are put off. *Note*: You can compute payments using an =If() function, but your instructor may instruct you to use the =Choose() function in some of the cells. *Hint*: Can you see that with a lag of 2, your January 2006 formula should yield zero for payments? With a lag of 3, are January and February payments both zero? (Answer: Yes, both are zero in that case.)
- INTEREST RATE: If the company can keep accounts payable at or less than $1,000,000, interest charged by the banker will be 7% annually. If accounts payable are more than $1,000,000 but less than $2,000,000, an annual interest rate of 9% will apply. If payables exceed $2,000,000, interest will be 12% annually.

CASH BUDGET Section

This is the "body" of the spreadsheet. Here the cash inflows and outflows are calculated and displayed, as shown in Figure 10-5. Only calculations through March 2006 are shown, but you should compute these amounts through December. An explanation of the line items follows the figure.

	A	B	C	D	E	F	G
31	**CASH BUDGET**	**OCT**	**NOV**	**DEC**	**JAN**	**FEB**	**MAR**
32	CASH RECEIPTS	**NA**	**NA**	**NA**			
33	CASH PAYMENTS						
34	PURCHASES	**NA**	**NA**	**NA**			
35	LABOR	**NA**	**NA**	**NA**			
36	ADMINISTRATIVE	**NA**	**NA**	**NA**			
37	ADVERTISING	**NA**	**NA**	**NA**			
38	INTEREST EXPENSE	**NA**	**NA**	**NA**			
39	TOTAL CASH PAYMENTS	**NA**	**NA**	**NA**			
40	CHANGE IN CASH (DECREASE)	**NA**	**NA**	**NA**			
41	CASH AT BEGINNING OF MONTH	**NA**	**NA**	**NA**	10000		
42	CASH AT BEGINNING OF MONTH PLUS CHANGE IN CASH	**NA**	**NA**	**NA**			
43	ADD: NEEDED BORROWING	**NA**	**NA**	**NA**			
44	CASH AT END OF MONTH	**NA**	**NA**	**NA**			

Figure 10-5 CASH BUDGET section

- CASH RECEIPTS: These are collections of receivables from the factor or customers. They have already been calculated.
- PURCHASES: Payments for raw materials purchases are a calculation.
- LABOR: Manufacturing labor is paid in the month goods are made at $2 per unit.
- ADMINISTRATIVE: Administrative expenses are a constant and can be echoed here.
- ADVERTISING: Advertising expenses are a constant and can be echoed here.

- INTEREST EXPENSE: In a month, this is a function of the debt owed to the bank at the beginning of the month and of the yearly interest rate, a calculation. Your formula should compute the amount for one month, of course.
- CHANGE IN CASH: The change in cash is total receipts less total payments.
- CASH AT BEGINNING OF MONTH: The cash at the beginning of a month equals the cash on hand at the end of the prior month. You should hard-code $10,000 for the beginning of January.
- The CHANGE IN CASH in a month, plus the CASH AT BEGINNING OF MONTH (cash on hand at the start of the month), gives the amount of cash before any borrowings at the end of a month.
- ADD: NEEDED BORROWING: If the previous amount is less than the desired minimum cash, the bank will lend enough money to get BigBiz's cash up to the minimum needed.
- CASH AT END OF MONTH: Cash on hand at the end of the month equals the change in cash during the month plus cash on hand at the beginning of the month plus any borrowing at the end of the month.

DEBT OWED Section

Figure 10-6 shows the calculation of debt owed at the end of a month. Only calculations through March 2006 are shown, but you should compute these amounts through December. An explanation of the line items follows the figure.

	A	B	C	D	E	F	G
46	**DEBT OWED**	**OCT**	**NOV**	**DEC**	**JAN**	**FEB**	**MAR**
47	DEBT OWED AT BEGINNING OF MONTH	**NA**	**NA**	**NA**	20000000		
48	ADD: NEEDED BORROWING	**NA**	**NA**	**NA**			
49	EQUALS: DEBT OWED AT MONTH END	**NA**	**NA**	**NA**			

Figure 10-6 DEBT OWED section

- DEBT OWED AT BEGINNING OF MONTH: Bank debt owed at the beginning of a month equals that owed at the end of the prior month. You should hard-code $20,000,000 for January.
- ADD: NEEDED BORROWING: Borrowings were computed in the **CASH BUDGET** section of the spreadsheet.
- EQUALS: DEBT OWED AT MONTH END: Debt owed at the end of a month equals debt owed at the beginning of the month plus any needed borrowings in the month.

ASSIGNMENT 2 USING THE SPREADSHEET FOR DECISION SUPPORT

You have built the spreadsheet model because the treasurer wants a better understanding of the advertising campaign results. The spreadsheet can now be used to answer questions that the treasurer has about the financial effects of the campaign.

The treasurer wants a way to see the financial impact of these decision variables:

- The advertising campaign's effect on units sold and selling prices
- Selling receivables to a factor
- Delaying payments to suppliers

Enter values for the factor and payment lag variables. In this way, you can see what strategy leaves the company in the best cash and debt position. Questions that you should be able to answer by using the Excel spreadsheet are set forth next. For each question, enter the inputs and then note the results on a separate sheet of paper.

- Assume that the company's suppliers would object strenuously to any delay in payments, and that the company would therefore maintain the current one-month lag. In that case, which strategy would be best—using a factor or not using a factor?
- Assume that the company's suppliers could be mollified by promises to renew normal payments in 2007. Which lag strategy would be best—a two-month lag or a three-month lag?
- Assume that the present policies for accounts payable and receivable continue. The advertising campaign will be instituted, but accounts receivable will not be sold to a factor, and having a one-month lag for accounts payable will prevail. Can this strategy be pursued without adding more debt at year-end?
- Looking at the summary of key results, is it clear from the outputs what strategy leads to the highest cash and lowest debt? If it's not clear what BigBiz should do, how should the spreadsheet be modified to aid decision-making?

Your instructor may have other questions for you to answer using the spreadsheet data. In addition, your instructor may ask that you chart some results.

When you are done with the spreadsheet, save it one last time (File—Save; **CASH.XLS** would be a good file name). Then, with the diskette in drive A:, use File—Close and then File—Exit.

You are now in a position to document your findings in a memorandum. Write a memorandum to the treasurer about your results. Observe the following requirements:

- Your memorandum should have a proper heading (DATE/ TO/ FROM/ SUBJECT). You may wish to use a Word memo template (File—New, click Memos, choose Contemporary Memo, and click OK).
- Briefly outline the ad campaign financing decision, including the questions that you have used the spreadsheet to answer. Then state the answers to those questions (and any others posed by your instructor).
- Supplement the text of your memo with any charts that your instructor requires.
- Supplement the text with a table summarizing results that you think are the most important. The form and the content of the table are up to you.

Enter your summary of data table into Word, using the following procedure:

1. Select the **Table** menu option, point to **Insert**, then click **Table**.
2. Enter the number of rows and columns.
3. Select **AutoFormat** and choose **Table Grid 1**.
4. Select **OK**, and then select **OK** again.

Assignment 3 Giving an Oral Presentation

Assume that the treasurer is so impressed with your memorandum that she has asked you to give an oral presentation of your findings to all those in BigBiz management. For your presentation, you'll want to include appropriate visual aids. Be prepared to defend your recommendations—management wants to be reassured about the worth of the ad campaign. Your presentation should take approximately 10 minutes, including a question-and-answer period.

DELIVERABLES

Assemble the following deliverables for your instructor:

1. Memorandum, including charts if required by your instructor
2. Presentation visual aids and handouts, as appropriate
3. Diskette, which should have your memorandum file and your Excel file

Staple your printouts together, with the memo on top. If the diskette holds more than just the file for this case, write a note to your instructor, stating the name of the **.xls** file.

The Merger Exchange-Ratio Analysis

11
CASE

DECISION SUPPORT USING EXCEL'S "WHAT IF?" CAPABILITY

PREVIEW

Company A wants to buy one of its competitors, Company B, through a friendly merger. First, shares of Company B's common stock would be exchanged for shares of Company A's common stock. Then, Company B's shares would be retired, and Company A would be the surviving company. The question is this: How many shares of Company A's stock should be given for each share of Company B's stock? In this case, you will use Excel to prepare a merged company income statement to help determine an acceptable common stock exchange ratio.

PREPARATION

- Review spreadsheet concepts discussed in class and/or in your textbook.
- Complete any exercises that your instructor assigns.
- Review any parts of Tutorials C or D that your instructor specifies, or refer to them as necessary.
- Review file-saving procedures for Windows programs. These are discussed in Tutorial D.
- Refer to Tutorial E as necessary.

BACKGROUND

A local manufacturer wants to buy one of its competitors. A regional bank will help conduct the acquisition, and the bank manager has called you in to help with modeling financial aspects of the merger. Your contact at the bank, Sue Green, wants to maintain confidentiality, and she will not reveal the company names. The acquiring company, you are told, is to be called Company A. The acquired company is to be called Company B.

The two companies have been competitors for many years. In recent years, Company B's profitability has dropped as the founder's health has declined. The founder wants to retire. The founder's children and relatives have no interest in running the company.

By contrast, Company A is a thriving business. Its managers are planning for expansion, and buying Company B is a good way of doing this. The management of both companies have agreed that the purchase will be stock-for-stock. New Company A common stock will be issued and given to Company B common stock shareholders. Company B common stock will be retired, and then only one company will remain, Company A, and only Company A stock will be held. The merged organization will be called Company A. All of Company B's assets will become part of Company A, and all of Company B's debts will be assumed by Company A. The management of Company A will run the combined organization.

The key question in everyone's mind is this: How many shares of Company A should be issued for each share of Company B? The two companies are roughly the same size and have about the same number of common shares outstanding. You tell Sue Green that you assume that one share of Company A's stock would be issued for one share of Company B's stock. You ask, "Stock certificates are just pieces of paper, aren't they?" Sue Green answers, "That sort of naiveté shows why I am the banker and you are the software specialist." Sue Green then gives you an overview of mergers and her clients' situation.

First, shares of Company A's stock are widely held and actively traded on stock exchanges. The shareholders of Company A want their common stock price to remain as high as possible. Shareholders would want any merger to enhance their common stock's value.

Sue Green says that financial professionals think that the price of a share of a company's common stock is based on the amount of the company's net income after taxes. In the long run, the higher a company's earnings, the higher the value of its common stock.

Earnings are usually expressed as earnings per share (EPS), which Sue Green tells you is a company's net income after taxes divided by the number of shares of common stock outstanding. ("Outstanding" means the number of shares owned by common stockholders.) Sue Green says that the EPS statistic expresses the amount of earnings that underlie each share of stock.

Sue Green gives you an example: Suppose that a company has $100 million in annual earnings and has 10 million shares outstanding. Its EPS would be $10—$100 million divided by 10 million. Assume that common stock investors would be willing to pay $100 per share in the stock market, Sue Green says. "That price would be 10 times the annual earnings per share—$100 price divided by $10 EPS," she points out.

"But then assume," says Sue Green, "that the company buys another company. This company has $50 million in earnings, and it also has 10 million shares outstanding. Its EPS is only $5—$50 million divided by 10 million. Assume further that the acquiring company issues another 10 million of its shares to buy the other company. So, now the acquiring company has $150 million in earnings and 20 million shares outstanding. Its EPS is now only $7.50, down from $10 before the merger. Thus, now shareholders have only $7.50 in EPS supporting each share of common stock, which was previously valued at $100. With lower EPS, the common stock's price would almost certainly fall, presumably to $75—10 times $7.50."

Sue Green concludes by saying, "Thus, the number of shares issued to buy a company has a huge impact on the EPS of the combined company. The buying company does not want to issue too many of its shares for fear of driving down its own post-merger stock price. The number of shares of its own stock that a company must issue to buy one share of the acquired company is called the *exchange ratio*. From the acquiring company's point of view, the lower the ratio the better."

To show her that you are not a complete financial neophyte, you ask why a company would buy another if its earnings prospects were not as good. "EPS is a ratio," you point out. "The denominator is the number of shares outstanding—what about the numerator, net income?"

Sue Green is impressed. "Very good! One company might buy another if managers see a situation that can be 'turned around,' and they think that they can make more profit using the company's assets than the current management is making. The idea that 2 plus 2 can equal 5, in mergers, is called *synergy*."

In the current situation, Sue Green explains, Company A's management thinks that the following Company B cost areas can be improved:

- Raw materials and labor are "variable costs" in manufacturing. The more units that are made, the higher the total variable cost. Company A's variable cost is $10 less per unit than Company B's cost per unit. Company A's costs are less simply because Company A's purchasing agents have negotiated better long-term sourcing agreements. In the future, all raw materials could be obtained under Company A's contract terms. Also, Company A's labor rates are somewhat lower than those of Company B for comparable jobs. In the future, all Company B's employees would be paid at Company A's rates. Jobs are at a premium. To keep their jobs, Company B's employees are expected to agree to lower wage rates.
- Company A pays its bankers 10% on its bank debt, while Company B must pay 11%. Company A would assume Company B's debts, of course. However, Company's A's treasurer says the company's bank will refinance Company B's debt at 10%. Thus, the merger would bring a one percentage point interest-rate saving on Company B's bank debt.
- Some of Company A's managers think that Company B's management is not very good. Fixed manufacturing costs—warehouse operations, salaries for plant managers, and so on—seem high. Many in Company A's management think they can bring that cost element down. Also, Company B's general corporate expenses—for marketing, for example—seem too high for the same reason and could be improved in the same way. This is where some of Company A's managers see the merger's "synergy effect." They will manage Company B's assets better than Company B's managers did, and the lower costs will flow through to higher net income after taxes and to higher EPS. Sue Green says, "Company A's managers think they are real smart, and maybe they are."

Sue Green notes that no one in Company A's management thinks that unit sales will improve because of the merger. All the improvements will be in the former Company B's costs.

Assume that it is the end of 2004. You must create a spreadsheet that (1) forecasts 2005 net income for both companies, (2) lets the user see the effects of improved management in the combined company, and (3) lets the user see the effects of different common stock exchange ratios on the merged company's EPS.

ASSIGNMENT 1 CREATING A SPREADSHEET FOR DECISION SUPPORT

In this assignment, you will produce a spreadsheet that models the merger exchange-ratio decision. In Assignments 2 and 3, you will use the spreadsheet model to develop information needed to determine the exchange ratio. In Assignment 2, you'll document your findings in a memorandum; in Assignment 3, you'll deliver those findings in an oral presentation. (*Note*: In what follows, references to Company B should be understood as shorthand for "former Company B operations in the merged company." After the merger, Company B itself will no longer exist, but its operations will continue within the larger Company A.)

In this assignment, you will create the 2005 spreadsheet in Excel. Your program will have *three* worksheets for the following:

- Forecast of Company A's 2005 income statement, standing alone. This sheet will have a CONSTANTS section and the forecasted INCOME STATEMENT section.
- Forecast of Company B's pre-merger 2005 income statement. This sheet will have a CONSTANTS section and the forecasted INCOME STATEMENT section. These are the results that Company A is buying, which Company A management thinks they can improve.
- Forecast of merged company financial results. Your sheet will have sections for constants, inputs, calculations, and the forecasted income statement for the combined company.

You will be shown how each sheet's sections should be set up before entering cell formulas. Your program will use Sheets 1, 2, and 3 in the spreadsheet program. A discussion of each sheet and each spreadsheet section follows. The discussion is about (1) how each section should be set up and (2) the logic of the formulas in the section's cells. When you type in the spreadsheet skeleton, follow the order given in this section.

Sheet for Company A's Income Statement

Your sheet will have sections for constants and the projected 2005 income statement. For convenience, you could use *Sheet2*.

CONSTANTS Section

Your sheet should have the constants shown in Figure 11-1. Line items are discussed in the comments following Figure 11-1.

	A	B	C	D	E
1	COMPANY A (ACQUIRING COMPANY) FINANCIAL RESULTS				
2	CONSTANTS		2005		
3	UNITS SOLD		60000		
4	SELLING PRICE / UNIT		200		
5	VARIABLE COST / UNIT		100		
6	FIXED MANUFACTURING COST		1000000		
7	ADMINISTRATIVE EXPENSE		1000000		
8	TAX RATE		0.4		
9	INTEREST RATE		0.1		
10	NUMBER OF SHARES OUTSTANDING		500000		
11	BEGINNING OF YEAR BANK DEBT		2500000		

Figure 11-1 CONSTANTS section

- UNITS SOLD and SELLING PRICE / UNIT: In 2005, Company A expects to sell 60,000 units of its product at $200 per unit.
- VARIABLE COST / UNIT: Raw materials and labor cost per unit are expected to be $100.
- FIXED MANUFACTURING COST: This cost will be $1 million, regardless of the production level.
- ADMINISTRATIVE EXPENSE: These general expenses will be $1 million in 2005.
- TAX RATE: The tax rate on pre-tax income will be 40%.
- INTEREST RATE: The interest rate on bank debt will be 10%. Interest will be paid on $2.5 million owed. No extra debt will be incurred in 2005.
- NUMBER OF SHARES OUTSTANDING: Company A has 500,000 shares of stock outstanding, pre-merger.
- BEGINNING OF YEAR BANK DEBT: The company owes $2,500,000 to start the year.

INCOME STATEMENT Section

Your sheet should have the forecasted income statement shown in Figure 11-2. Line items are discussed in the comments following Figure 11-2.

	A	B	C
14	**INCOME STATEMENT**		**2005**
15	SALES		
16	VARIABLE COSTS		
17	FIXED MANUFACTURING COST		
18	ADMINISTRATIVE EXPENSE		
19	TOTAL COSTS		
20	INCOME BEFORE INTEREST AND TAXES		
21	INTEREST EXPENSE		
22	NET INCOME BEFORE TAXES		
23	INCOME TAX EXPENSE		
24	NET INCOME AFTER TAXES		

Figure 11-2 INCOME STATEMENT section

- SALES: Sales are a function of the number of units sold and the unit selling price.
- VARIABLE COSTS: Variable costs are a function of the number of units sold and the variable cost per unit.
- FIXED MANUFACTURING COST and ADMINISTRATIVE EXPENSE are constants that can be echoed here.
- TOTAL COSTS are the sum of variable costs, fixed manufacturing cost, and administrative expense.
- INCOME BEFORE INTEREST AND TAXES: This is total sales less total costs.
- INTEREST EXPENSE: This is a function of the interest rate and the bank debt at the beginning of the year.
- NET INCOME BEFORE TAXES: This is income before interest and taxes less interest expense.
- INCOME TAX EXPENSE: Income taxes are zero if income before taxes is less than or equal to zero; otherwise, apply the tax rate to income before taxes.
- NET INCOME AFTER TAXES: Net income after taxes is net income before taxes less income tax expense.

Sheet for Company B's Income Statement

Your sheet will have sections for constants and the projected 2005 pre-merger income statement. For convenience, you could use *Sheet3*.

CONSTANTS Section

Your sheet should have the constants shown in Figure 11-3. Line items are defined in the same way as for Company A.

	A	B	C	D	E
1	COMPANY B (ACQUIRED COMPANY) FINANCIAL RESULTS				
2	CONSTANTS		2005		
3	UNITS SOLD		50000		
4	SELLING PRICE / UNIT		200		
5	VARIABLE COST / UNIT		110		
6	FIXED MANUFACTURING COST		1100000		
7	ADMINISTRATIVE EXPENSE		1200000		
8	TAX RATE		0.4		
9	INTEREST RATE		0.11		
10	NUMBER OF SHARES OUTSTANDING		550000		
11	BEGINNING OF YEAR BANK DEBT		3000000		

Figure 11-3 CONSTANTS section

INCOME STATEMENT Section

Your sheet for Company B should have a forecasted INCOME STATEMENT section in the same format as shown in Figure 11-2. Line items are defined in the same way as for Company A.

Sheet for Merged Company Financial Results

Your sheet should have sections for constants, inputs, calculations, and results, and a combined income statement. This sheet will summarize data from the other two sheets, and you could use *Sheet1* for it.

CONSTANTS Section

Your sheet should have the constants shown in Figure 11-4. These data are from the company income statement sheets, and they will be needed again in the combined company income statement.

	A	B	C	D
1	MERGED COMPANY 2005 FINANCIAL RESULTS			
2	CONSTANTS		A	B
3	NUMBER OF SHARES OUTSTANDING BEFORE MERGER		500000	550000
4	INTEREST RATE		0.10	0.11
5	TAX RATE		0.4	0.4

Figure 11-4 CONSTANTS section

- NUMBER OF SHARES OUTSTANDING BEFORE MERGER: Company A has 500,000 shares outstanding, and Company B has 550,000, pre-merger.
- INTEREST RATE: Company A pays 10% on its bank debt; Company B pays 11%.
- TAX RATE: The tax rate for both companies is 40%.

INPUTS Section

Your spreadsheet should have the inputs shown in Figure 11-5. You change the inputs in order to play "what if" with the analysis.

	A	B	C
7	**INPUTS**		
8	SYNERGY COEFFICIENT		
9	EXCHANGE RATIO: # OF SHARES OF A FOR ONE SHARE OF B		

Figure 11-5 INPUTS section

- SYNERGY COEFFICIENT: The synergy coefficient denotes the degree of improvement in Company B fixed manufacturing and administrative operations that can be expected in the merged company. Example: If a user enters .10, then Company B's fixed manufacturing and administrative expenses would then be 10% less in the merged company than if there were no merger.
- EXCHANGE RATIO: # OF SHARES OF A FOR ONE SHARE OF B: Enter the number of shares Company A issues to retire a share of Company B. If 1 is entered, the exchange ratio is 1:1. If .5 is entered, the exchange ratio is one-half of a share of A for each share of B (perhaps more clearly, this would work out to 1 share of A for 2 shares of B, but .5 would nevertheless be the entry here).

The income statement also has cells that allow the user to make choices. These inputs are described in the **INCOME STATEMENT** section.

CALCULATIONS / RESULTS Section

Figure 11-6 shows amounts that should be calculated, including the key calculation, merged EPS, which is the result that Company A owners and management are most concerned with knowing. Calculations are based on constants, other calculations, and/or amounts in the **INCOME STATEMENT** section. Cells with "NA" should have no calculation. An explanation of the line items follows Figure 11-6.

	A	B	C	D	E
11	CALCULATIONS / RESULTS		A	B	MERGED
12	EARNINGS PER SHARE				
13	# OF A SHARES ISSUED FOR B SHARES			NA	NA
14	TOTAL MERGED SHARES			NA	NA

Figure 11-6 CALCULATIONS / RESULTS section

- EARNINGS PER SHARE: Company A and B earnings per share are a function of the pre-merger net income after taxes for each company and their respective shares outstanding. Merged company earnings per share are a function of combined net income

Case 11

after taxes and total merged shares. (*Note*: Net income after tax amounts are shown in the 2005 Income Statement, discussed next.)

- # OF A SHARES ISSUED FOR B SHARES: The number of Company A's shares issued for Company B's shares is a function of the exchange ratio (an input) and the number of Company B's shares (a constant).
- TOTAL MERGED SHARES: Total merged shares is the sum of Company A's shares already outstanding (a constant) and the shares issued to buy Company B (a calculation).

The key result in this case is the MERGED earnings per share, shown highlighted. You can achieve the highlighting by using Format—Cells—Border; select the Outline and the Line Style.

INCOME STATEMENT Section

This is the "body" of the spreadsheet. Company A and Company B income statements are echoed here. An adjusted Company B income statement is calculated to reflect merger efficiencies. A combined post-merger income statement is computed. Figure 11-7 shows the form of the income statement. An explanation of the line items follows the figure.

An amount in one sheet can be shown in another sheet using "exclamation point" notation. Example: Assume that in Sheet1, cell B20, you want to show the value that is in Sheet2, cell C30. In Sheet1, cell B20, you would have this formula: =Sheet2!C30.

Similarly, an amount in one sheet can be used in a formula in another sheet. Example: In Sheet1, cell B20, you want to show the product of values in Sheet2, cell C30, and Sheet3, cell F21. In Sheet1, cell B20, you would have this formula: =Sheet2!C30 * Sheet3!F21.

	A	B	C	D	E	F
17		EXPECT				
18		BETTER B			B	
19		RESULTS	A	B	ADJUSTED	A + B
20		IF MERGED	NO	NO	FOR	MERGED
21	2005 INCOME STATEMENT	(1=YES)	MERGER	MERGER	MERGER	COMPANY
22	SALES	0				
23	VARIABLE COSTS	1				
24	FIXED MANUFACTURING	1				
25	ADMINISTRATIVE EXPENSE	1				
26	TOTAL COSTS	NA				
27	INCOME BEFORE INTEREST AND TAXES	NA				
28	INTEREST EXPENSE	1				
29	INCOME BEFORE TAXES	NA				
30	INCOME TAX EXPENSE	NA				
31	NET INCOME AFTER TAXES	NA				

Figure 11-7 INCOME STATEMENT section

- The EXPECT BETTER B RESULTS IF MERGED column lets the user indicate if a Company B amount should improve with the merger. Example: If the user thinks that Company B's variable costs will be better with the merger, then a **1** is entered; otherwise, a **0** is entered. In the figure's example, no sales improvement is expected, but all four major cost elements are expected to improve. Note that subtotals and other amounts are not applicable (NA) for this purpose.
- The A NO MERGER and B NO MERGER column values are the Company A and Company B forecasted income statement results, echoed from their respective sheets.

- The B ADJUSTED FOR MERGER column shows Company B's amounts, adjusted for the effects of the merger. For example, assuming the user expects better Company B results, variable costs would be shown here computed using Company A's variable cost per unit. Assuming that the user expects improvement, interest expense would be shown here computed using Company A's interest rate. Assuming the user expects improvement, fixed manufacturing and administrative expenses would be shown here, reduced by the synergy effect. The amount of the synergy effect would be a function of the synergy coefficient input.
- A + B MERGED COMPANY results are Company A's results plus adjusted Company B's results.

ASSIGNMENT 2 USING THE SPREADSHEET FOR DECISION SUPPORT

You have built the spreadsheet model because Sue Green and Company A's management want a way to compute an appropriate exchange ratio—one that will preserve or improve the EPS of the combined company. The spreadsheet can now be used to answer questions from Sue Green and Company A's management. They want a way to see the financial impact of these decision variables:

- The assumed impact of better Company A management on Company B's operations—the synergy effect and other improvements
- The EPS impact of different exchange ratios
- The impact of individual cost elements on the decision

Enter values for synergy and for the exchange ratio and also indicate cost elements that are expected to improve in the income statement. In this way, you can show the impact on merged company earnings per share.

Questions that you should be able to answer by using the Excel spreadsheet are set forth next. For each question, enter the inputs and then make a note of the results on a separate sheet of paper.

- Assume a synergy effect of 5%, which seems very reasonable to most of Company A's managers. Assume that all four Company B cost elements will improve. What is the exchange ratio needed to preserve earnings per share at the same level in the merged company?
- A minority of Company A's managers wonder about the reliability of the synergy effect for fixed manufacturing and administrative expenses. They think it might be unwise to count on this. They want to assume a synergy effect of 0%, but they want to retain the assumption that variable cost and interest-rate benefits will occur. Given those assumptions, what is the exchange ratio needed to preserve earnings per share in the merged company?
- Assume that Company B's owners say they could not possibly part with their shares unless they received at least .75 a share of Company A for each share of Company B. What synergy effect would be needed to preserve earnings per share in the merged company with that .75 exchange ratio? (The question assumes that variable cost and interest-rate benefits will occur.)

Your instructor may have other questions he or she wants answered using the spreadsheet. In addition, your instructor may ask that you chart some results.

When you are done with the spreadsheet, save it one last time (File—Save; **MERGE_AB.XLS** would be a good file name). Then, with the diskette in drive A:, use File—Close and then select File—Exit.

You are now in a position to document your findings in a memorandum. Write a memorandum to Sue Green that states your results. Observe the following requirements:

- Your memorandum should have a proper heading (DATE/ TO/ FROM/ SUBJECT). You may wish to use a Word memo template (File—New, click Memos, choose Contemporary Memo, and click OK).
- Briefly outline the exchange-ratio decision, including the questions that you have used the spreadsheet to answer. Then state the answers to those questions (and any others posed by your instructor).
- Enter your summary data into a Word table that summarizes the results that you think are most important. The form and content of the table are up to you. To make a table in Word, use the following procedure:
    1. Select the **Table** menu option, click **Insert**, and then click **Table**.
    2. Enter the number of rows and columns.
    3. Select **AutoFormat** and the desired format—**Table Grid 1** is a good choice.
    4. Select **OK** and then select **OK** again.
- Supplement the text of your memo with any charts that your instructor requires.

Assignment 3 Giving an Oral Presentation

Assume that Sue Green is so impressed by your memorandum that she has asked you to give an oral presentation of your findings to all those in Company A's management. For your presentation, you'll want to include appropriate visual aids and handouts. Be prepared to defend your recommendations—management wants to be reassured about the financial wisdom of buying Company B. Your presentation should take approximately 10 minutes, including a question-and-answer period.

Deliverables

Assemble the following deliverables for your instructor:

1. Memorandum, including charts if required by your instructor
2. Presentation visual aids and handouts, as appropriate
3. Diskette, which should have your memorandum file and your Excel file

Staple your printouts together, with the memo on top. If the diskette holds more than just the file for this case, write a note to your instructor, stating the name of the **.xls** file.

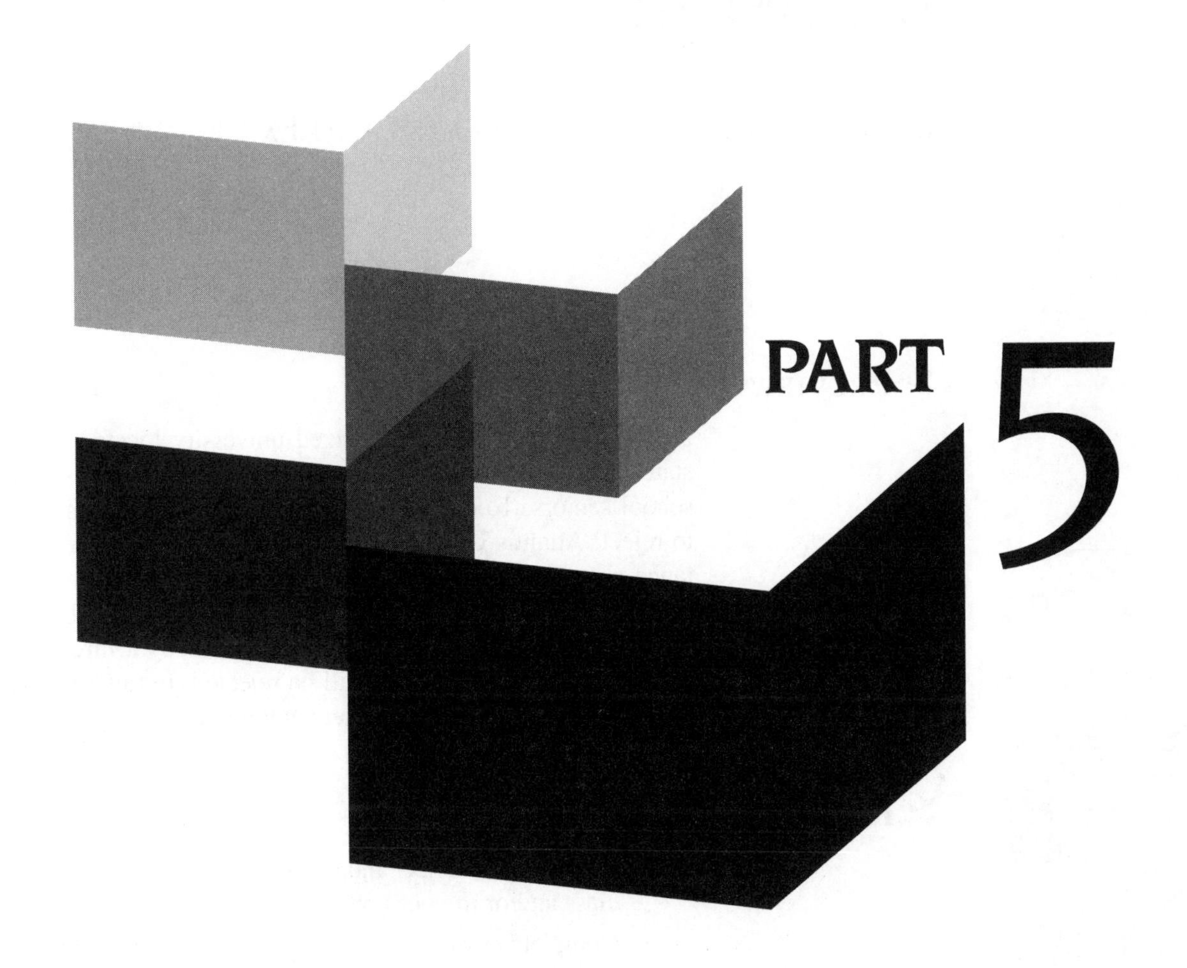

Integration Case Using Access and Excel

12 CASE

Atlantis University's Applicant-Acceptance Decision

DECISION SUPPORT USING ACCESS AND EXCEL

PREVIEW

Atlantis University is a mid-sized university located in a nearby state. Each year, Atlantis is flooded with applications from high school seniors. To decide whom to accept as freshmen and whom to reject, Atlantis University officials want to calculate an "acceptance score" for each applicant. The score is based on an applicant's junior- and senior-year high school GPAs, SAT score, and other factors. Applicants whose score exceeds a threshold value will be accepted, and others will be rejected. In this case, you will use both Access and Excel to help automate the scoring.

PREPARATION

- Review spreadsheet and database concepts discussed in class and/or in your textbook.
- Complete any exercises that your instructor assigns.
- Obtain the database file **ADMIT.mdb** from your instructor.
- Review any part of Tutorials A, B, C, or D that your instructor specifies, or refer to them as necessary.
- Review file-saving procedures for Windows programs. These are discussed in Tutorial D.
- Refer to Tutorial E as necessary.

BACKGROUND

Each year, Atlantis University gets thousands of applications from high school seniors wanting to attend the university. Until now, an army of university officials has reviewed each application thoroughly—but unsystematically. The time-consuming review process has resulted in some odd acceptance decisions; in many cases, it has been difficult to explain why one high school student was accepted and another with a similar academic record was rejected.

Fearing lawsuits, university officials have decided to adopt a scoring system for the incoming freshman class. Officials believe that a numerical system keyed to certain objective factors will be the most defensible. Applicants will be assigned points for high school GPA, SAT score, and other factors. Points will be accumulated into a total score. Applicants whose score exceeds a certain threshold will be accepted, and others will be rejected. In this case, you will use both Access and Excel to create the new scoring system.

Here are the criteria to be used in calculating the acceptance score:

- Junior year GPA
- Senior year GPA
- Trend in GPA (senior year GPA increasing or decreasing from junior year)
- SAT score
- Residence status (in-state applicants are favored over out-of-state residents)

As applications are received in the U.S. mail and online, university clerks enter selected data into an Access database. The database is called **ADMIT.mdb**. Its tables have admissions data about each applicant. The **ADMIT.mdb** tables are explained next. (*Note*: To continue, you will need to obtain **ADMIT.mdb** from your instructor.)

APPLICANT Table

This table shows the student I.D. number, student name, and residence status for each applicant. The Student Number field is the table's primary key field. Partial data are shown in Figure 12-1. "In State?" is a Yes/No field. The check mark (for Yes) indicates that the applicant is an in-state resident.

APPLICANT : Table

Student Number	Name	In State?
1001	Smith	☐
1002	Jones	☑
1003	Albers	☐
1004	Ruth	☑

Figure 12-1 APPLICANT table (partial data)

SAT SCORES Table

This table shows the student's I.D. number, date(s) on which the student took the SAT, and the SAT score. The score is the total out of 1,600 possible points. The primary key is the compound key formed by the Student Number and the Test Date fields. Partial data are shown in Figure 12-2.

Student Number	Test Date	Score
1001	30-Sep-03	1000
1001	15-Dec-03	1100
1001	15-Sep-04	1200
1002	30-Sep-03	1200
1002	15-Sep-04	1100

Figure 12-2 SAT SCORES table (partial data)

JUNIOR GRADES Table

This table records data about academic courses the applicant took in his or her junior year—the name of the course and the letter grade. (Non-academic courses such as driver's ed and gym are not recorded. To simplify the analysis, Atlantis drops plusses and minuses from transcript grades.) The primary key is the compound key formed by the Student Number and the Course fields. Partial data are shown in Figure 12-3.

Student Number	Year	Course	Grade
1001	JR	ACCOUNTING 1	B
1001	JR	ALGEBRA	C
1001	JR	ART 1	B
1001	JR	BIOLOGY	B
1001	JR	COMPUTER 1	A
1001	JR	DRAFTING	B
1001	JR	ENGLISH 1	B
1001	JR	ENGLISH 2	D
1001	JR	HEALTH	C
1001	JR	HISTORY 1	A
1001	JR	ROTC	A
1001	JR	SOCIAL STUDIES	C
1002	JR	AGRICULTURE 1	B
1002	JR	ALGEBRA	D
1002	JR	ART 1	B
1002	JR	CHEMISTRY	A

Figure 12-3 JUNIOR GRADES table (partial data)

There is an analogous SENIOR GRADES table that shows the courses and grades that an applicant has taken in his or her senior year. The structure of the table is the same as the JUNIOR GRADES table.

ASSIGNMENT 1 USING ACCESS AND EXCEL FOR DECISION SUPPORT

Assume that it is late in the academic year, and that Atlantis administrators need to process applications for the next school year. Administrators need an automated way to process applicant data. You must use Access and Excel to help Atlantis calculate scores for each applicant. Your work should result in a summary Excel worksheet that (1) calculates acceptance points for the five factors (junior year GPA, senior year GPA, trend in GPA, SAT score, and residence status) and (2) provides an accept/reject recommendation for each applicant.

The suggested format for this summary worksheet is shown in Figure 12-4. The calculations are discussed after the figure (data are illustrative).

	A	B	C	D	E	F	G	H	I
1	CALCULATION OF ADMITTANCE POINTS AND ACCEPTANCE DECISION								
2	NAME	STUDENT #	JR GPA	SR GPA	IN STATE	SAT	GPA TREND	TOTAL	ACCEPT?
3	Smith	1001	11.333	15.417	0	34	5	65.750	ACCEPT
4	Jones	1002	10.545	14.167	5	34	5	68.712	ACCEPT
5	Albers	1003	11.636	15.000	0	36	5	67.636	ACCEPT
6	Ruth	1004	10.800	19.167	5	35	5	74.967	ACCEPT

Figure 12-4 Calculation of admittance points and acceptance decision (illustrative data)

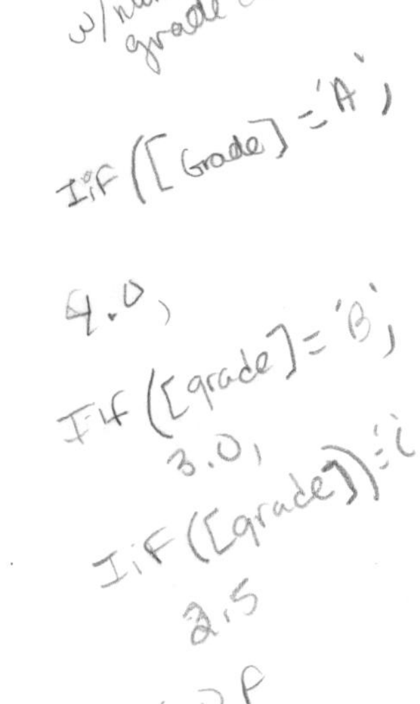

- JR GPA points equal four times the junior year GPA (calculated elsewhere in the spreadsheet file).
- SR GPA points equal five times the senior year GPA (calculated elsewhere).
- IN STATE applicants are awarded five points. Out-of-state residents are not awarded points. In-state status is shown elsewhere in the spreadsheet.
- SAT points equaling the integer value of the square root of the SAT score are awarded. (Use the =INT() and =SQRT() functions appropriately.) The SAT score is shown elsewhere in the spreadsheet.
- GPA TREND is calculated thusly: Five points are awarded if the senior year GPA exceeds the junior year GPA, five points are deducted if the senior year GPA is less than the junior year GPA, and no points are awarded if the two GPAs are the same. GPAs are shown elsewhere in the spreadsheet.
- TOTAL points are the sum of points for the five factors.
- ACCEPT? reflects the threshold value, which is 58. If total points exceed 58, accept the applicant; otherwise, reject the applicant.

Your spreadsheet file will need other worksheets that feed data to the summary worksheet. Your strategy for developing these other worksheets will include the following:

- Using Access queries to generate appropriate data
- Importing Access query output and Access table data into Excel
- Using Excel to generate information for the summary worksheet

You can rely on the following Access and Excel techniques:

- Running Sigma queries in Access
- Using pivot tables in Excel (see Tutorial E; your instructor may have further guidance)
- Importing data into Excel; the sequence of steps is as follows:
  1. Select Data—Import External Data—Import Data.
  2. Specify the Access **.mdb** file in the window.

3. Select the Table or Query in the window.
4. Specify the spreadsheet cell location. Click OK.

- Moving data between worksheets using SheetX! cell reference notation (example: assume that you want to put the value of Sheet 2's cell C4 into cell K2 of Sheet 1; in cell K2 you would put this formula: =Sheet2!C4)
- Inserting a blank worksheet (Insert—Worksheet)

Recommended Access Queries

You should make certain queries in Access. Query output can then be imported into Excel for your analysis. The recommended queries are discussed next.

Query: Count of Junior Courses By Student

The query should compute the number of courses the applicant took in his or her junior year. Partial output is shown in Figure 12-5.

Count Of Junior Courses By Student : Select Query

Student Number	Count Of Junior Courses
1001	12
1002	11
1003	11
1004	10
1005	10

Figure 12-5 Query: Count of Junior Courses By Student (partial data)

You should make a similar query that counts the number of courses taken in the senior year for each student.

Query: Maximum SAT By Student

A student can take the SAT any number of times. Atlantis wants to award points for only the best performance. You should make a query that calculates the maximum SAT score for each student. Partial data is shown in Figure 12-6.

Max SAT By Student : Select Query

Student Number	Max SAT Score
1001	1200
1002	1200
1003	1300
1004	1250
1005	910

Figure 12-6 Query: Maximum SAT By Student (partial data)

Additional Excel Worksheets

Your summary worksheet should pull data from other worksheets in the **.xls** file. These other worksheets are discussed next.

Calculation of Junior Credits Earned in Courses

You need a worksheet that computes grade points and credits earned by each student in the junior year. Partial data is shown in Figure 12-7. Line items are discussed following the figure. You should make a similar worksheet for senior grades.

	A	B	C	D	E	F
1	StuNum	Year	Course	Grade	GradePoints	Credits
2	1001	JR	BIOLOGY	B	3	9
3	1001	JR	ALGEBRA	C	2	6
4	1001	JR	HISTORY 1	A	4	12
5	1001	JR	ENGLISH 1	B	3	9
6	1001	JR	DRAFTING	B	3	9
7	1001	JR	SOCIAL STUDIES	C	2	6
8	1001	JR	ENGLISH 2	D	1	3
9	1002	JR	CHEMISTRY	A	4	12
10	1002	JR	ALGEBRA	D	1	3

Figure 12-7 Calculation of junior credits earned in courses (partial data)

- StuNum (Student Number), Year, Course, and Grade data are available in an Access table. (Row 1 headers will be different when data are imported from Access. The headers can be changed to save space, as was done for Figure 12-7.)
- GradePoints are calculated as follows: Four grade points are earned for an A, 3 for a B, 2 for a C, and 1 for a D. None is earned for an F.
- Credits for courses are earned by credit hour; courses are assumed to be 3 credit-hour courses, so Credits earned are 3 times GradePoints earned.

Pivot Table for Total Junior Year Credits Earned

You should make a pivot table in Excel that adds up credits earned by each student in his or her junior year. Partial data is shown in Figure 12-8. Pivot table data can be pulled from the calculation of credits worksheet. You should make a similar pivot table for students' work in their senior year.

Case 12

	A	B
1		
2	Junior Credits	
3	Sum of Credits	
4	StuNum	Total
5	1001	102
6	1002	87

Figure 12-8 Pivot table for total junior year credits earned (partial data)

Calculation of GPAs

You need a worksheet that computes junior and senior year GPAs and that holds other data. Partial data is shown in Figure 12-9. A discussion of the data follows the figure.

	A	B	C	D	E	F	G	H	I	J
1	StuNum	Name	InState?	SATScore	NumberOf JrCourses	NumberOf SrCourses	Junior Credits	JrGPA	Senior Credits	SrGPA
2	1001	Smith	FALSE	1200	12	12	102	2.833	111	3.083
3	1002	Jones	TRUE	1200	11	12	87	2.636	102	2.833
4	1003	Albers	FALSE	1300	11	12	96	2.909	108	3.000
5	1004	Ruth	TRUE	1250	10	12	81	2.700	138	3.833

Figure 12-9 Calculation of GPAs (partial data)

- StuNum, Name, InState?, and SATScore data are available from Access tables.
- Data for the number of junior and senior courses are available from Access query output. (Row 1 headers will be different when data are imported from Access. The headers can be changed to save space, as was done for Figure 12-9.)
- Junior and senior credits are available in pivot tables.
- GPAs are based on credits earned and the number of courses (each course is assumed to be three hours).

Other Required Calculations

In addition to the calculations previously noted, you will need other computations when you document your results in a memorandum, as follows:

- Compute the number of students accepted and the number rejected. (*Hint*: Use the CountIF() function to do this.)
- Compute the number of in-state and out-of-state students accepted.
- Compute the average senior year GPA and SAT scores for those accepted.

You will need to decide where to put these calculations in your worksheet and how to do them. (The output for these calculations is not shown in the figures.) Next, in Assignments 2 and 3, you must perform calculations and then document the results of using the Accept/Reject program. In Assignment 2, you will write a memorandum to Atlantis University's president about the results. In Assignment 3, you will prepare an oral presentation. Guidance on your Access and Excel work is given next.

Assignment 2 Using the Spreadsheet for Decision Support

You have used Access and Excel to automate Atlantis' Accept/Reject decision. Finish creating your spreadsheet. When you are done with the spreadsheet, save the file one last time (File—Save). A good file name would be **ADMITS.xls**. Then, with the diskette in drive A:, use File—Close and then File—Exit.

You are now in a position to document your work in a memorandum. Write a memorandum to the university president about your work. Observe the following requirements:

- Your memorandum should have a proper heading (DATE/ TO/ FROM/ SUBJECT). You may wish to use a Word memo template (File—New, under Templates in the New Document Task Pane click Memos, choose Contemporary Memo).
- Outline the purpose of the Accept/Reject program and briefly state how it works.
- Tell the president how many students applied, how many were accepted, and how many were rejected. Indicate how many in-state and out-of-state students were accepted. Provide the senior year GPA and SAT score for those students who were accepted.

ASSIGNMENT 3 GIVING AN ORAL PRESENTATION

Assume that the president is so impressed by your memorandum that she has asked you to give an oral presentation of your findings to Atlantis University's Board of Trustees. For your presentation, you'll want to include appropriate visual aids and handouts. Your presentation should take approximately 10-15 minutes, possibly longer, including a question-and-answer period.

DELIVERABLES

Assemble the following deliverables for your instructor:

1. Memorandum
2. Presentation visual aids and handouts, as appropriate
3. Diskette, which should have your memorandum file, your Access file, and your Excel file

Staple your printouts together, with the memo on top. If your diskette holds database and spreadsheet files other than for this case, write a note to your instructor, stating the names of the **.xls**, **.mdb**, and **.doc** files to be graded.

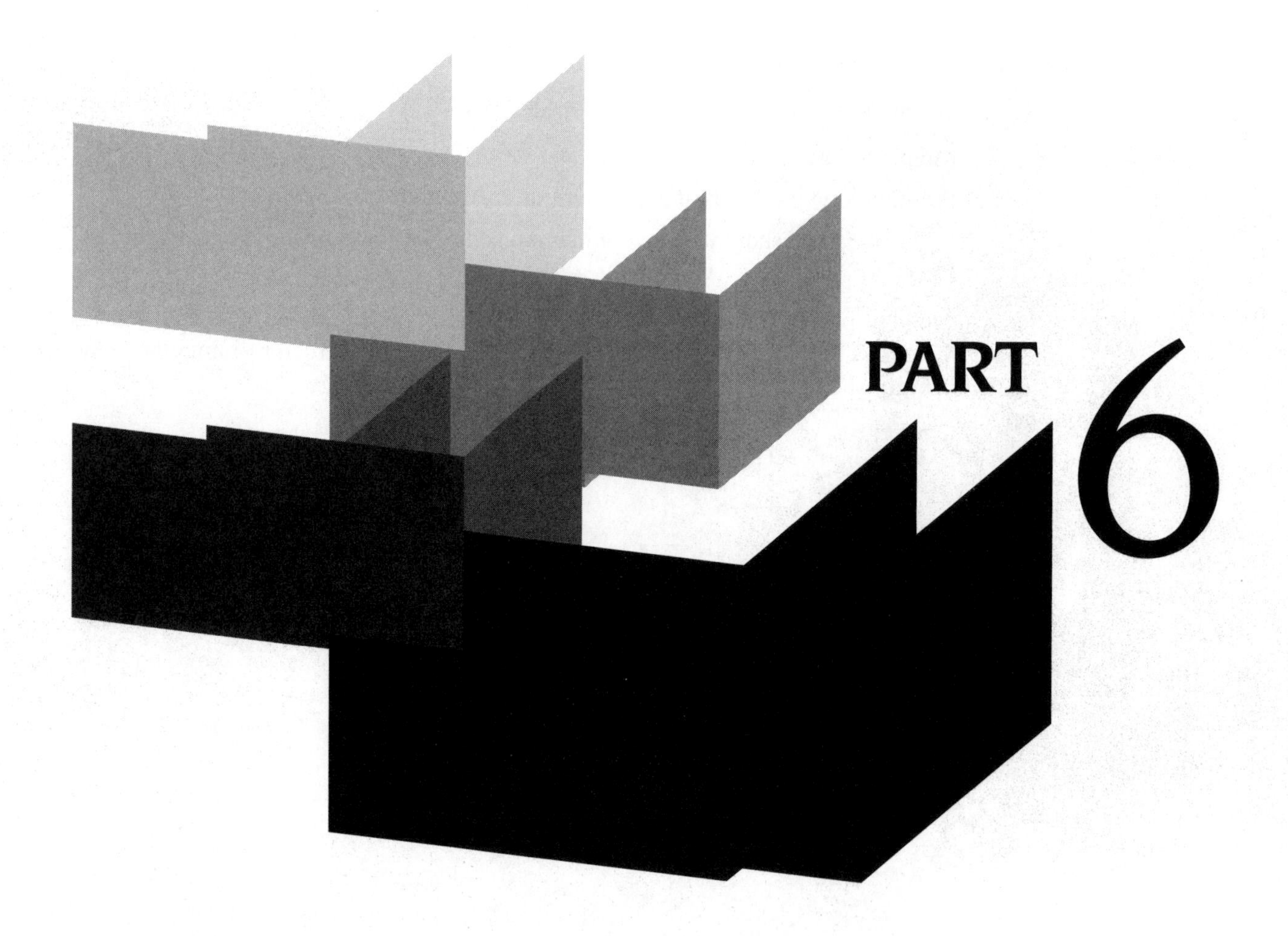

Presentation Skills

Giving an Oral Presentation

Giving an oral presentation provides you with the opportunity to practice the presentation skills you'll need in the workplace. The presentations you'll create for the cases in this textbook will be similar to real-world presentations: You'll present objective, technical results to an organization's stakeholders. You'll support your presentation with visual aids commonly used in the business world. Your instructor may wish to have your classmates role-play an audience of business managers, bankers, or employees and have them give you feedback on your presentation.

Follow these four steps to create an effective presentation:

1. Plan your presentation.
2. Draft your presentation.
3. Create graphics and other visual aids.
4. Practice your delivery.

Let's start at the beginning and look at the steps involved in planning your presentation.

PLAN YOUR PRESENTATION

When planning an oral presentation, you'll need to know your time limits, establish your purpose, analyze your audience, and gather information. Let's look at each of these elements.

Know Your Time Limits

You'll need to consider your time limits on two levels. First, consider how much time you'll have to deliver your presentation. What can you expect to accomplish in 10 minutes? The element of time is the "driver" of any presentation. It limits the breadth and depth of your talk—and the number of visual aids that you can use. Second, consider how much time you'll need for the actual process of preparing your presentation: drafting your presentation, creating graphics, and practicing your delivery.

Establish Your Purpose

You must define your purpose: what you need and want to say and to whom. For the cases in the Access portion of the book, your purpose will be to inform and explain. For instance, a business' owners, managers, and employees need to know how their organization's database is organized and how to use it to fill in input forms, create reports, and so on. By contrast, for the cases in the Excel portion of the book, your purpose will be to recommend a course of action. You'll be making recommendations based on your results from inputting various scenarios to business owners, managers, and bankers.

Analyze Your Audience

Before drafting your presentation, analyze your audience. Ask yourself these questions: What does my audience already know about the subject? What do they want to know? What do they need to know? Do they have any biases that I should consider? What level of technical detail is best suited to their level of knowledge and interest?

In some Access cases, you will make a presentation to an audience who might not be familiar with Access or databases in general. In other cases, you might be giving a presentation to a business owner who started work on the database but was not able to finish it. Tailor your presentation to suit your audience.

For the Excel cases, you will be interpreting results for an audience of bankers and business managers. The audience does not need to know the detailed technical aspects of how you generated your results. What they *do* need to know is what assumptions you made prior to developing your spreadsheet because those assumptions might have an impact on their opinion of your results.

Gather Information

Since you will have just completed a case, you'll have the basic information. For the Access cases, review the main points of the case and your goals. Be sure to include all the points that you feel are important for the audience. In addition, you may wish to go beyond the requirements and explain additional ways in which the database could be used to benefit the organization, now or in the future.

For the Excel cases, you can refer to the tutorials for assistance in interpreting the results from your spreadsheet analysis. For some cases, you might want to research the Internet for business trends or background information that could be used to support your presentation. For example, for Case 9, The New Wave Mutual Fund Investment Mix Problem, you may want to do Internet research so you can more confidently predict the future of the economy and the financial marketplace.

Drafting Your Presentation

You might be tempted to write your presentation and then memorize it, word for word. If you do, your presentation will sound very unnatural because when people speak, they use a simpler vocabulary and shorter sentences than when they write. Thus, you may want to draft your presentation by noting just key phrases and statistics. When drafting your presentation, follow this sequence:

1. Write the main body of your presentation.
2. Write the introduction to your presentation.
3. Write the conclusion to your presentation.

Writing the Main Body

When you draft your presentation, write the body first. If you try to write the opening paragraph first, you'll spend an inordinate amount of time creating "the perfect paragraph"—only to revise it after you've written the body of your presentation.

Keep Your Audience in Mind

To write the main body, review your purpose and your audience's profile. What are the main points you need to make? What are your audience's wants, needs, interests, and technical expertise? It's important to include some basic technical details in your presentation, but keep in mind the technical expertise of your audience.

What if your audience consists of people with different needs, interests, and levels of technical expertise? For example, in the Access cases, an employee might want to know how to input information into a form, but the business owner might already know how to input data and will be more interested in generating queries and reports. You'll need to acknowledge their differences in your presentation. Thus, you might want to say something like, "And now, let's look at how data entry clerks can input data into the form."

Similarly, in the Excel cases, your audience will usually consist of business owners, managers, and bankers. The owners' and managers' concerns will be profitability and growth. By contrast, the bankers' main concern will be getting a loan repaid. You'll need to address the interests of each group.

Use Transitions and Repetition

Because your audience can't read the text of your presentation, you'll need to use transitions to compensate. Words such as *next*, *first*, *second*, *finally*, etc., will help your audience follow the sequence of your ideas. Words such as *however*, *by contrast*, *on the other hand*, and *similarly* will help them to follow shifts in thought. You can also use your voice and hand gestures to convey emphasis.

Also think about how you can use body language to emphasize what you're saying. For instance, if you are stating three reasons, you can tick them off on your fingers as you discuss them: one, two, three. Similarly, if you're saying that profits will be flat, you can make a level motion with your hand for emphasis.

As you draft your presentation, repeat key points to emphasize them. For example, suppose that your point is that outsourcing labor will provide the greatest gains in net income. Begin by previewing that concept: State that you're going to demonstrate how outsourcing labor will yield the greatest profits. Then provide statistics that support your claim and show visual aids that graphically illustrate your point. Summarize by repeating your point: "As you can see, outsourcing labor does yield the greatest profits."

Rely on Graphics to Support Your Talk

As you write the main body, think of how you can best incorporate graphics into your presentation. Don't waste a lot of words describing what you're presenting if you can use a graphic that can quickly portray it. For instance, instead of describing how information from a query is input into a report, show a sample, a query result, and a completed report. Figure E-1 and E-2 illustrate this.

Query for Report : Select Query

Last Name	First Name	Date	Type	Amount Paid
Angerstein	Amy	6/30/2003	Catalina 23	30
Angerstein	Amy	7/2/2003	Catalina 23	52.5
Codfish	Colburn	7/5/2003	Capri 14.2	72
Dellabella	Debra	7/5/2003	Hobie 18	28
Edelweiss	Edna	7/11/2003	Sunfish 12	150
Angerstein	Amy	8/1/2003	Row	36
Babbles	Bob	6/29/2003	Capri 14.2	24
Babbles	Bob	6/30/2003	Capri 14.2	24
Dellabella	Debra	7/7/2003	Hobie 18	14
Edelweiss	Edna	7/12/2003	Sunfish 12	25
Codfish	Colburn	7/11/2003	Capri 14.2	24

Figure E-1 Access query

Customer Bill

Last Name	*First Name*	*Date*	*Type*	*Amount Paid*
Angerstein	*Amy*			
		8/1/2003	Row	$36.00
		7/2/2003	Catalina 23	$52.50
		6/30/2003	Catalina 23	$30.00
	Total			*$118.50*
Babbles	*Bob*			
		6/30/2003	Capri 14.2	$24.00
		6/29/2003	Capri 14.2	$24.00
	Total			*$48.00*

Figure E-2 Access report

Also consider what kinds of graphics media are available—and how well you know how to use them. For example, if you've never prepared a PowerPoint presentation, will you have enough time to learn how to do it before your presentation?

Anticipate the Unexpected

Even though you're just drafting your report now, eventually you'll be answering audience questions. Being able to handle questions smoothly is the mark of a professional. The first step is being prepared for those questions.

You won't use all the facts you gather in your presentation. However, as you draft your presentation, you might want to keep some of those facts jotted on paper—just in case you need them to answer questions from the audience. For instance, for some Excel presentations you might be asked why you are not recommending some course of action that you did not mention in your report.

Writing the Introduction

After you have written the main body of your talk, then develop an introduction. An introduction should be only a paragraph or two in length and preview the main points that your presentation will cover.

For some of the Access cases, you might want to include some general information about databases: what they can do, why they are used, and how they can help the company become more efficient and profitable. You won't need to say much about the business operation since the audience already works for the company.

For the Excel cases, you might want to have an introduction to the general business scenario and describe any assumptions you made when creating and running your decision support spreadsheet. Excel is used for decision support, so describe the choices and decision criteria.

Writing the Conclusion

Every good presentation needs a good ending. Don't leave the audience hanging! Your conclusion should be brief—only a paragraph or two in length—and give your presentation a sense of closure. Use the conclusion to repeat your main points or, for the Excel cases, your findings and/or recommendations.

CREATING GRAPHICS

Using visual aids is a powerful method of getting your point across and making it understandable to your audience. Visual aids come in a variety of physical forms. Some forms are more effective than others.

Choosing Graphics Media

The media that you use should depend on your situation and what media are available. One of the key things to remember when using any media is this: *You must maintain control of the media or you'll lose control of your audience.*

The following list highlights some of the most common media and their strengths and weaknesses.

- **Handouts:** This medium is readily available in both classrooms and businesses. It relieves the audience from taking notes. Graphics can be in full color, of professional quality, and multi-colored. *Negatives*: You must stop and take time to hand out individual sheets. During your presentation, the audience might be studying and discussing your handouts rather than listening to you. Lack of media control is *the* major drawback—and it can kill your presentation.
- **Chalkboard:** This informal medium is readily available in the classroom but not in many businesses. *Negatives*: You'll need to turn your back on the audience when you're writing (thus losing control of them), and you'll need to erase what you've written as you go. Your handwriting must be very good. In addition, attractive graphics are difficult to create.
- **Flip Chart:** This informal medium is readily available in many businesses. *Negatives*: The writing space is so small, it's not effective for more than a very small group. This medium shares many of the same negatives as the chalkboard.

Tutorial E

- **Overheads:** This medium is readily available in both classrooms and businesses. You do have control over what the audience sees and when. You can create very professional PowerPoint presentations on overhead transparencies. *Negatives*: Handwritten overheads look amateurish. Without special equipment and preparation, graphics are difficult to do well.
- **Slides:** This formal medium is readily available in many businesses and can be used in large rooms. Slides can be either 35mm slides or the more popular electronic on-screen slides, which is usually *the* medium of choice for most large organizations. It is slick and professional and is generally preferred for formal presentations. *Negatives*: You must have access to the equipment and know how to use it. It takes time to learn how to create and use computer graphics. Also, you must have some source of ambient light, or it will be difficult to see your notes in the dark.

Creating Charts and Graphs

Technically, charts and graphs are not the same thing, although many graphs are called "charts." Usually, charts show relationships, and graphs show change. However, Excel makes no distinction and calls both charts.

Charts are easy to create in Excel. Unfortunately, they are so easy to create that people often create graphics that are meaningless or that inaccurately reflect the data they represent. Let's look at how to select the most appropriate graphics.

Charts

Use pie charts to display data that is related to a whole. Excel takes the numbers you want to graph and makes them a percentage of 100. You might use a pie chart when showing the percentage of shoppers who bought a generic brand of toothpaste versus a major brand, as shown in Figure E-3. You would *not*, however, use a pie chart to show a company's net income over a three-year period, because the period cannot be considered "a whole" or the years its "parts," as shown in Figure E-4.

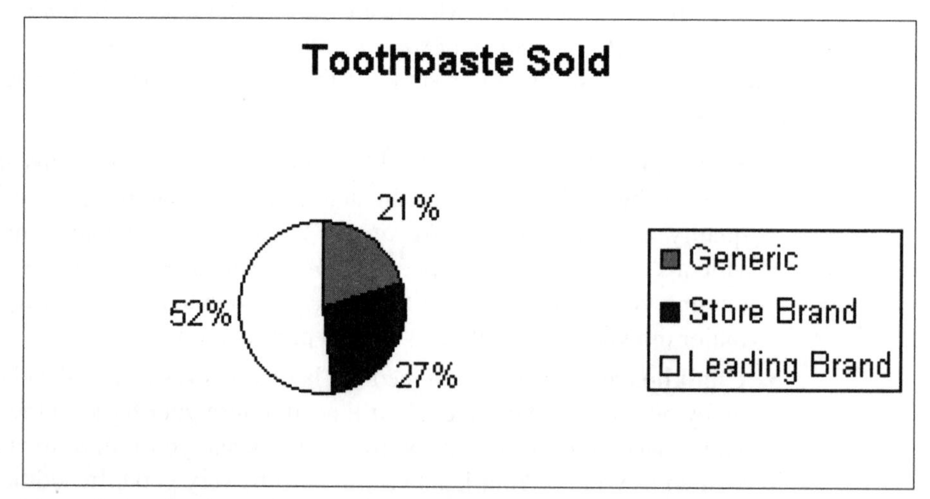

Figure E-3 Pie chart: Appropriate use

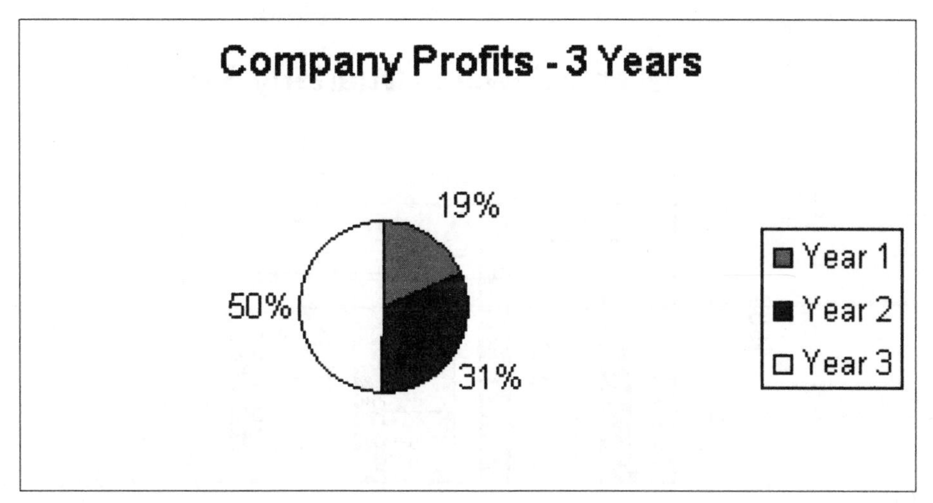

Figure E-4 Pie chart: Inappropriate use

Use bar charts when you want to compare several amounts at one time. For example, you might want to compare the net profit that would result from each of several different strategies. You can also use a bar chart to show changes over time. For example, you might show how one pricing strategy would increase profits year after year.

When you are showing a graphic, don't forget that you need labels that explain what the graphic shows. For instance, if you're showing a graph with an X and Y axis, you should show what each axis represents so the audience doesn't puzzle over the graphic while you're speaking. Figures E-5 and E-6 show the necessity of labels.

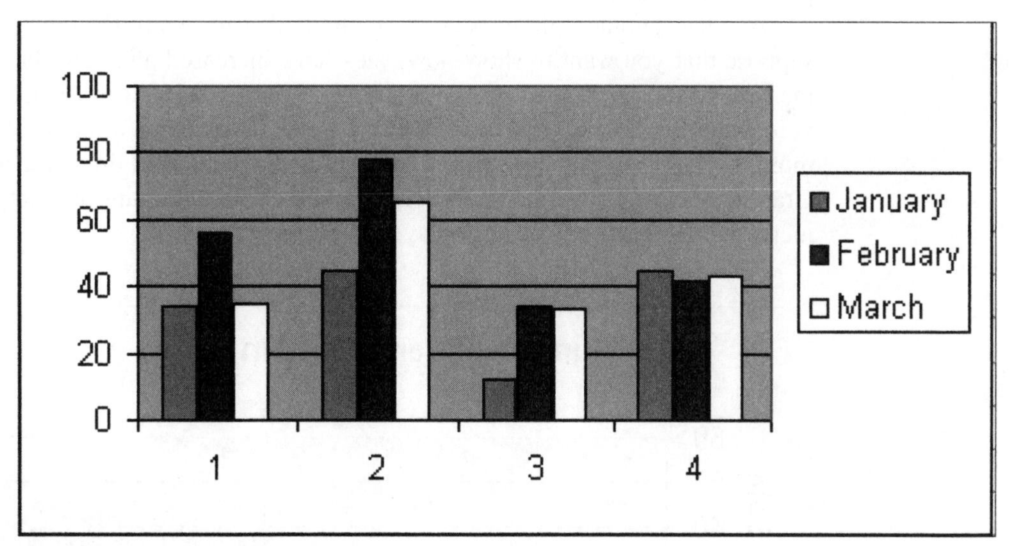

Figure E-5 Graphic without labels

Tutorial E

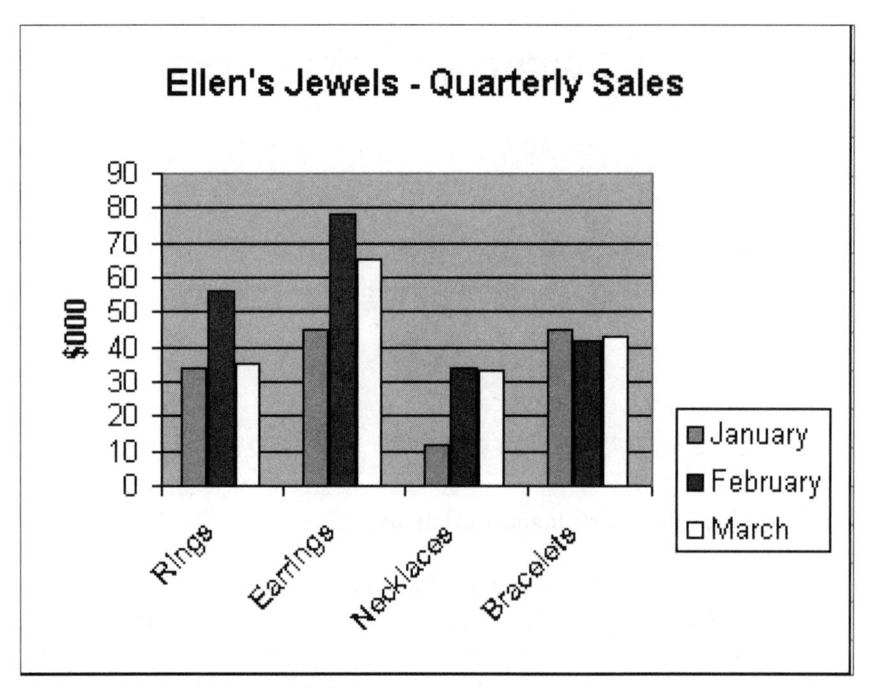

Figure E-6 Graphic with labels

In Figure E-5, the graphic is not labeled, and neither are the X and Y axes: Are the amounts shown units or dollars? What elements are represented by each bar? By contrast, Figure E-6 provides a comprehensive snapshot of the business operation—which would support a talk rather than distract from it.

Another common pitfall is creating charts that have a misleading premise. For example, suppose that you want to show how sales have increased and contributed to a growth in net income. If you graph the number of items sold, as displayed in Figure E-7, it might not tell you about the actual dollar value of those items; however, it might be more appropriate (and more revealing) to graph the profit margin for the items sold times the number of items sold. Graphing the profit margin would give a more accurate picture of what is contributing to the increased net income. This is displayed in Figure E-8.

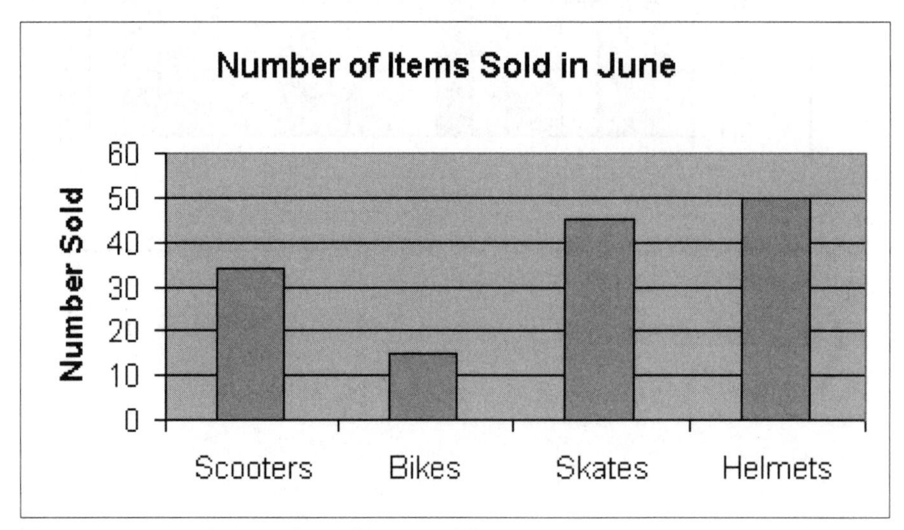

Figure E-7 Graph: Number of items sold

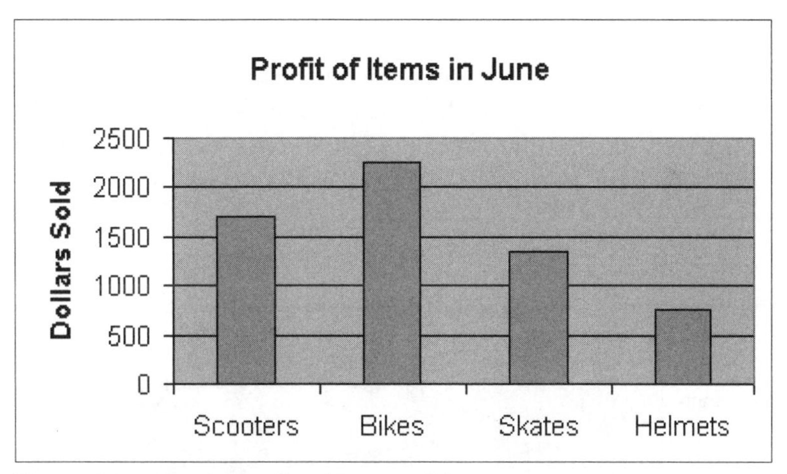

Figure E-8 Graph: Profit of items sold

Another common pitfall is putting too much data in a single, comparative chart. Here is an example: Assume that you want to compare monthly mortgage payments for two loans with different interest rates and timeframes. You have a spreadsheet that computes the payment data, shown in Figure E-9.

Calculation of Monthly Payment						
Rate	6.00%	6.10%	6.20%	6.30%	6.40%	6.50%
Amount	100000	100000	100000	100000	100000	100000
Payment (360 payments)	$599	$605	$612	$618	$625	$632
Payment (180 payments)	$843	$849	$854	$860	$865	$871
Amount	150000	150000	150000	150000	150000	150000
Payment (360 payments)	$899	$908	$918	$928	$938	$948
Payment (180 payments)	$1,265	$1,273	$1,282	$1,290	$1,298	$1,306

Figure E-9 Calculation of monthly payment

You try to capture all this information in a single Excel chart, such as the one shown in Figure E-10.

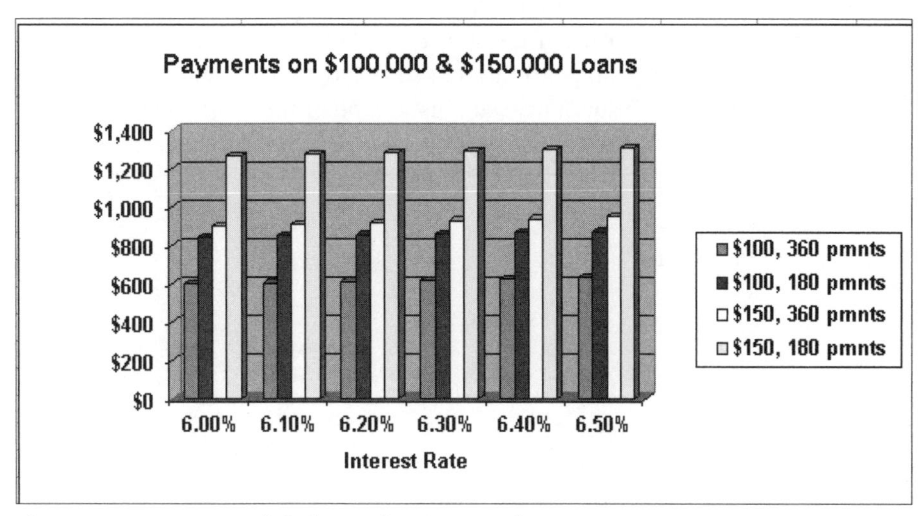

Figure E-10 Too much information in one chart

There is a great deal of information here. Most readers would probably appreciate it if you broke things up a bit. It would probably be easier to understand the data if you made one chart for the $100,000 loan and another one for the $150,000 loan. The chart for the $100,000 loan would look like the chart shown in Figure E-11.

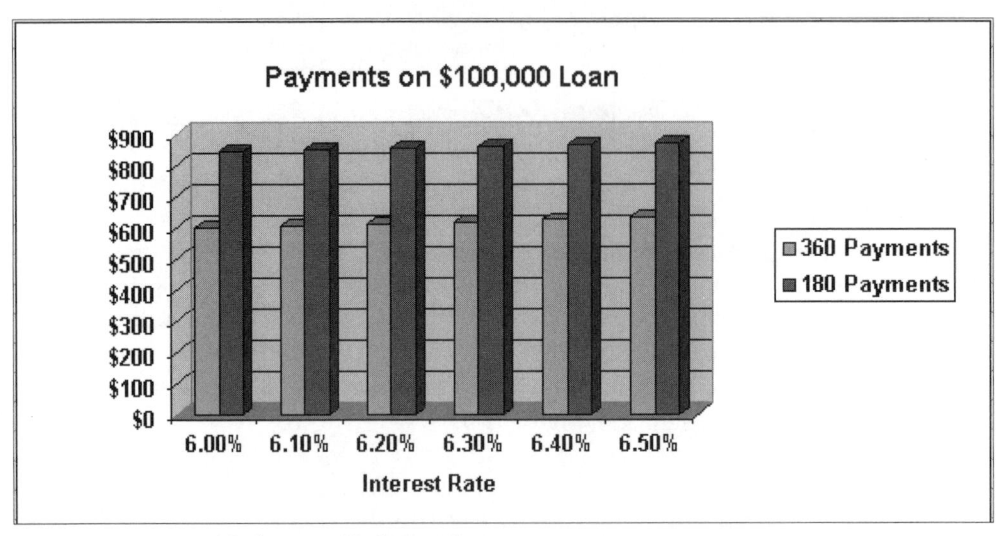

Figure E-11 Good balance of information

A similar chart could be made for the $150,000 loan. The charts could then be augmented by text that summarizes the main differences between the payments for each loan. In this fashion, the reader is led step by step through the data analysis.

You may wish to use the Chart Wizard in Excel, but be aware that the Charting functions can be tricky to use at times, especially with sophisticated charting. Some tweaking to the chart is often necessary. Your instructor may be able to provide specific directions for your individual charts.

Making a Pivot Table in Excel

Suppose that you have data for a company's sales transactions by month, by salesperson, and by amount for each product type. You would like to display each salesperson's total sales, according to type of product sold and also by month. Using a pivot table in Excel, you can tabulate such summary data, using one or more dimensions.

Figure E-12 shows total sales, cross-tabulated by salesperson and by month. This display was created by using a pivot table in Excel.

	A	B	C	D	E
1	**Name**	**Product**	**January**	**February**	**March**
2	Jones	Product 1	30,000	35,000	40,000
3	Jones	Product 2	33,000	34,000	45,000
4	Jones	Product 3	24,000	30,000	42,000
5	Smith	Product 1	40,000	38,000	36,000
6	Smith	Product 2	41,000	37,000	38,000
7	Smith	Product 3	39,000	50,000	33,000
8	Bonds	Product 1	25,000	26,000	25,000
9	Bonds	Product 2	22,000	25,000	24,000
10	Bonds	Product 3	19,000	20,000	19,000
11	Ruth	Product 1	44,000	42,000	33,000
12	Ruth	Product 2	45,000	40,000	30,000
13	Ruth	Product 3	50,000	52,000	35,000

Figure E-12 Excel spreadsheet data

You can create this kind of table (and many other kinds) with Excel's Pivot Table tool. You can use the following steps to create a pivot table.

1. Select Data—Pivot Table and PivotChart Report. You will see the screen shown in Figure E-13.

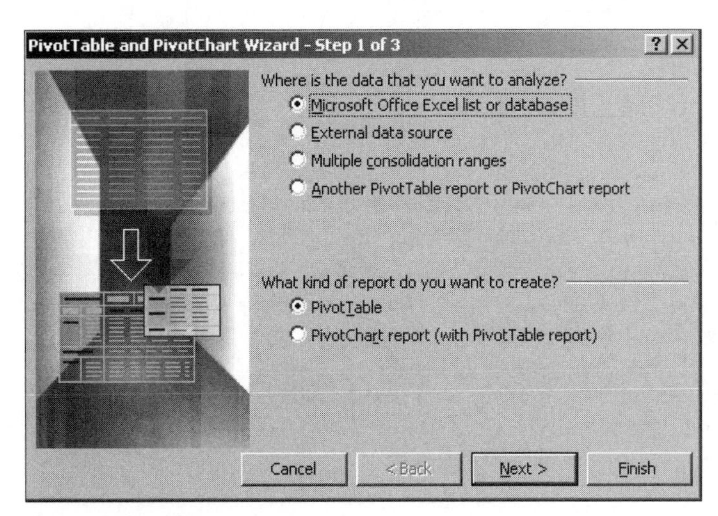

Figure E-13 Step 1

2. To make a pivot table, click Next. You will see the screen shown in Figure E-14. By default the range will be the most northeast contiguous data range in the spreadsheet. You can change this in the Range window.

Figure E-14 Step 2

3. Click Next. You will then see the screen shown in Figure E-15.

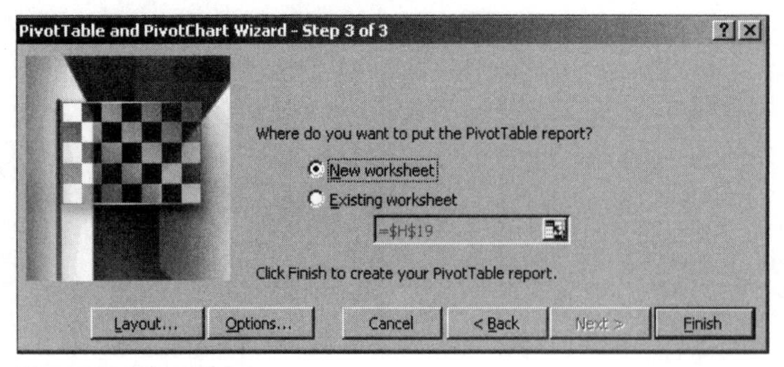

Figure E-15 Step 3

4. You can put the table in the current sheet (probably "Sheet1") or in a separate sheet. The latter way is shown here. *New worksheet* is the default. Click Finish.
5. You will see the screen shown in Figure E-16. The data range's column headings are shown in the Pivot Table Field List. Click and drag column headings into the Row, Column, and Data areas.

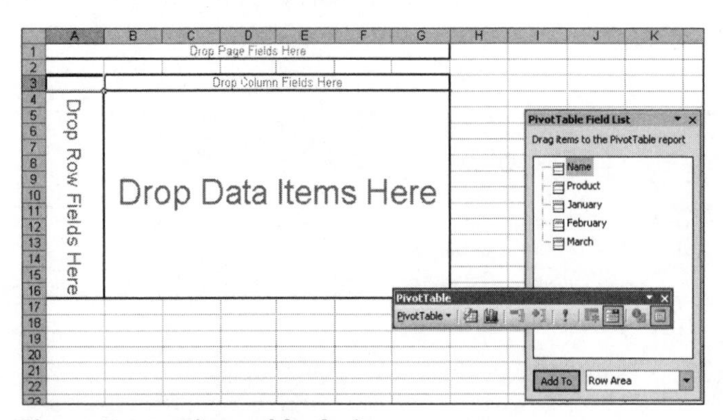

Figure E-16 Pivot table design

6. Assume that you want to see the total sales, by product, for each salesperson. You would drag the Name field to the "Drop Column Fields Here" area, and you should see the result shown in Figure E-17.

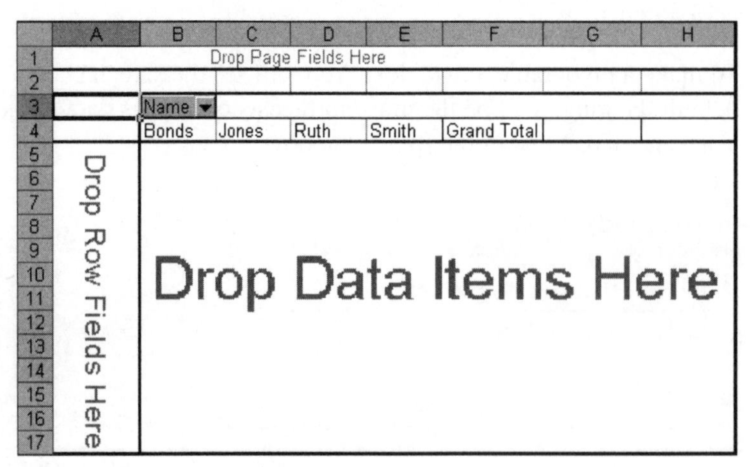

Figure E-17 Column fields

7. Next, take the Product field and drag it to the "Drop Row Fields Here" area, and you should see the result shown in Figure E-18.

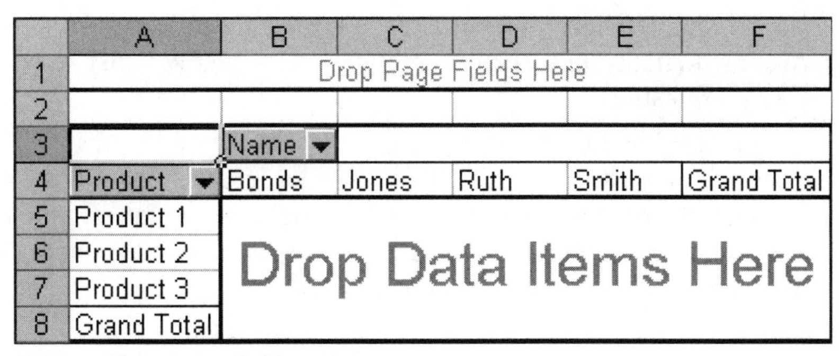

	A	B	C	D	E	F
1	Drop Page Fields Here					
2						
3		Name				
4	Product	Bonds	Jones	Ruth	Smith	Grand Total
5	Product 1	Drop Data Items Here				
6	Product 2					
7	Product 3					
8	Grand Total					

Figure E-18 Row fields

8. Finally, you would take the month fields (January, February, and March) and drag them individually to the "Drop Data Items Here" area to produce the final pivot table, and you should see the result shown in Figure E-19.

	A	B	C	D	E	F	G
1							
2							
3			Name				
4	Product	Data	Bonds	Jones	Ruth	Smith	Grand Total
5	Product 1	Sum of January	25000	30000	44000	40000	139000
6		Sum of February	26000	35000	42000	38000	141000
7		Sum of March	25000	40000	33000	36000	134000
8	Product 2	Sum of January	22000	33000	45000	41000	141000
9		Sum of February	25000	34000	40000	37000	136000
10		Sum of March	24000	45000	30000	38000	137000
11	Product 3	Sum of January	19000	24000	50000	39000	132000
12		Sum of February	20000	30000	52000	50000	152000
13		Sum of March	19000	42000	35000	33000	129000
14	Total Sum of January		66000	87000	139000	120000	412000
15	Total Sum of February		71000	99000	134000	125000	429000
16	Total Sum of March		68000	127000	98000	107000	400000

Figure E-19 Data items

By default, Excel adds up all the sales for each salesperson by month for each individual product. It also shows the total sales for each month for all products at the bottom of the pivot table.

Creating PowerPoint Presentations

PowerPoint presentations are easy to create: Simply open up the application and use the appropriate slide layout for a title slide, a slide containing a bulleted list, a picture, a graphic, and so on. In choosing a design template (the background color, the font color and size, and the fill-in colors for all slides in your presentation), keep these guidelines in mind:

- Avoid using pastel background colors. Dark backgrounds such as blue, black, and purple work well on overhead projection systems.

- If your projection area is small or your audience is large, you might want to use boldface type for all your text to make it even more visible.
- Try using "transition" slides to keep your talk lively. A variety of styles are in the program and available for use. Common transitions include "dissolves" and "wipes." Avoid wild transitions, such as swirling letters, that will distract your audience from your presentation.
- You can use "build" effects if you do not want your audience to see the whole slide when you show it. A "build" effect will allow each bullet to come up when the mouse button or the right arrow is clicked. A "build" effect allows you to control the visual and explain the elements as you go. This can be controlled under the Custom Animation screen, as shown in Figure E-20.

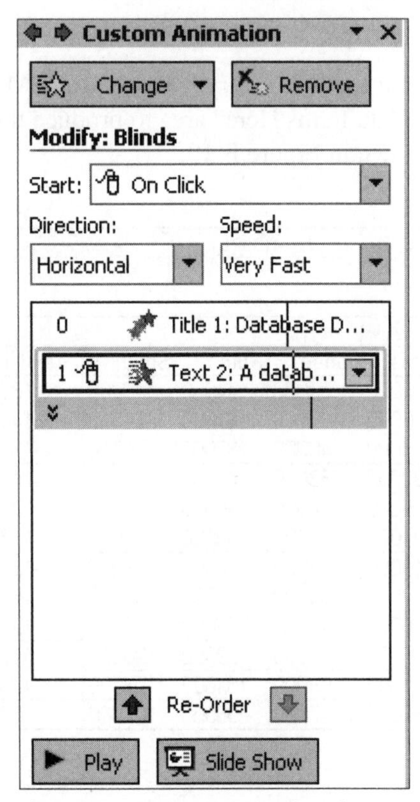

Figure E-20 Custom Animation screen

- You can create PowerPoint slides that have a section for notes. These are printed for the speaker when you choose the drop-down menu on the File—Print dialog box, as shown in Figure E-21. Each slide is printed as half-size, with the notes written underneath each slide, as shown in Figure E-22.

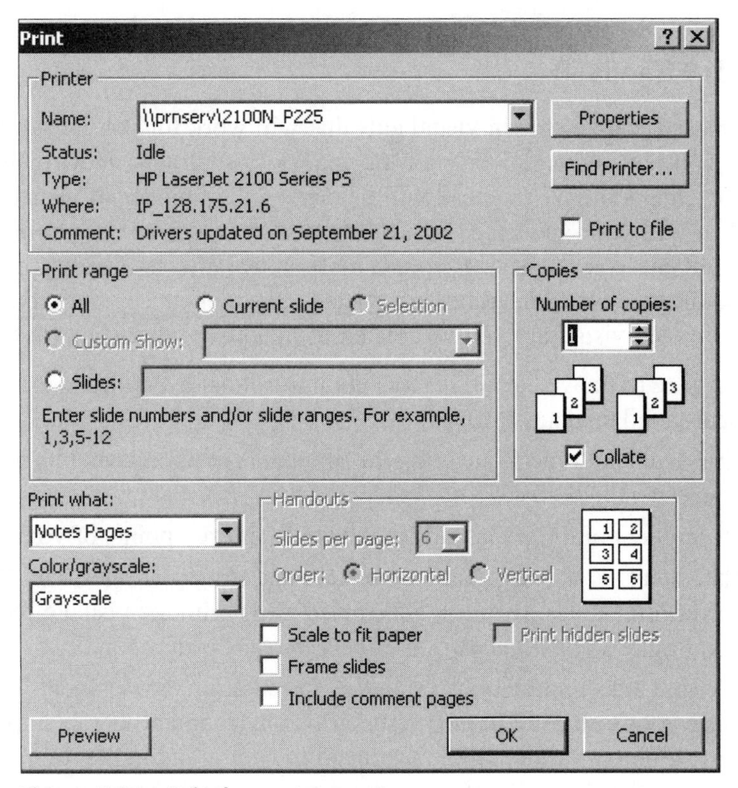

Figure E-21 Printing notes page

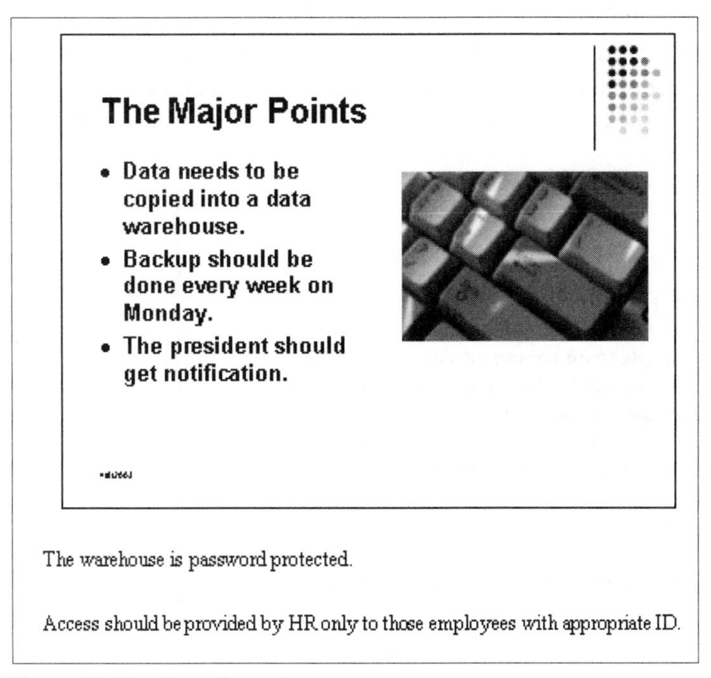

Figure E-22 Sample notes page

- As previously mentioned, always check your presentation on the overhead. What looks good on your computer screen might not be readable on an overhead screen.

Using Visual Aids Effectively

Make sure that you've chosen visual aids that will work for you most effectively. Also make sure that you have enough—but not too many—visual aids. How many is too many? The amount of time you have to speak will determine the number of visual aids that you should use, and so will your audience. For example, if you will be addressing a group of teenage summer helpers, you might want to use more visual effects than if you make a presentation to a board of directors. Remember, use visual aids to enhance your talk, not replace it.

Review each visual aid you've created to make sure that it meets the following criteria:

- The size of the visual aid is large enough so that everyone in the audience can see it clearly and read any labels.
- The visual aid is accurate, e.g., the graphics are not misleading and there are no typos or misspelled words.
- The content of the visual aid is relevant to the key points of your presentation.
- The visual aid doesn't distract the audience from your message. Often when creating PowerPoint slides, speakers get carried away with the visual effects, e.g., they use spiraling text and other jarring effects. Keep it professional.
- A visual aid should look good in the presentation environment. If at all possible, try using your visual aid in the presentation environment. For example, when using PowerPoint, try it out on the overhead projector and in the room in which you'll be showing the slides. What looks good on your computer screen might not look good on the overhead projector when viewed from a distance of 20 feet.
- Make sure that all numbers are rounded unless decimals or pennies are crucial.
- Do not make your slides too busy or crowded. Most experts say that bulleted lists should contain no more than four or five lines. Also avoid having too many labels. A busy slide is illustrated in Figure E-23.

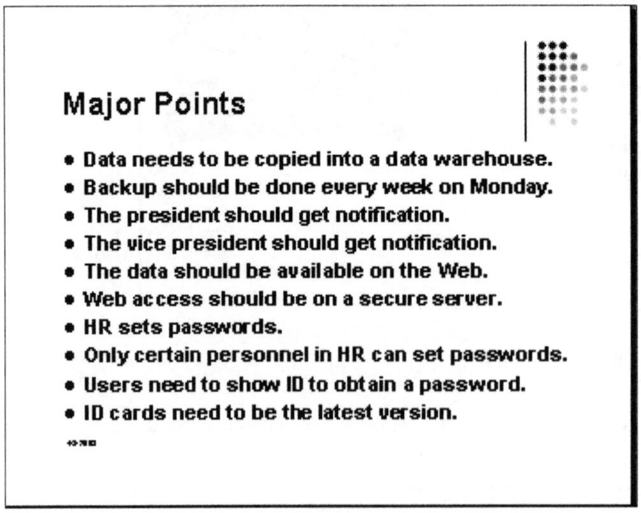

Major Points

- Data needs to be copied into a data warehouse.
- Backup should be done every week on Monday.
- The president should get notification.
- The vice president should get notification.
- The data should be available on the Web.
- Web access should be on a secure server.
- HR sets passwords.
- Only certain personnel in HR can set passwords.
- Users need to show ID to obtain a password.
- ID cards need to be the latest version.

Figure E-23 Busy slide

PRACTICING FOR YOUR DELIVERY

Surveys indicate that public speaking is most people's greatest fear. However, fear or nervousness can be a positive factor. It can channel your energy into doing a good job. Remember that an audience will rarely perceive that you are nervous unless you fidget or your voice cracks. They are present to hear the content of your talk, so think of the audience, not how you feel.

The presentations you give for the cases in this textbook will be in a classroom setting with 20 to 40 students. Ask yourself this question: Would I be afraid to talk to just one or two of my classmates? Think of your presentation as an extended conversation with several of your classmates. Let your gaze shift from person to person and make eye contact with them. As your gaze drifts around the room, say to yourself, "I'm speaking to one person." As you become more experienced in speaking before groups, you will be able to let your gaze move naturally from one audience member to another.

Tips for Practicing Your Delivery

Giving an effective presentation is not reading a report to an audience. Rather, it requires that you have your message rehearsed well enough so you can present it naturally, confidently, and in tandem with well-chosen visual aids. Make sure that you allow sufficient time to practice your delivery.

- Practice your presentation several times and use your visual aids when you practice.
- Show visual aids at the right time and only at the right time. A visual aid should not be shown too soon or too late. In your speaker's notes, you might even have cues for when to show each visual aid.
- Maintain eye and voice contact with the audience when using the visual aid. Don't look at the screen or turn your back on the audience.
- Use your visual aids and refer to them both in your talk and with hand gestures. Don't ignore your own visual aid.
- Keep in mind that your visual aids should support your presentation, not *be* the presentation. In other words, don't have everything you are going to say on each slide. Use visual aids to illustrate the key points and statistics and fill in with your talk.
- Time check: Are you within time limits?
- Using numbers effectively: Use round numbers when speaking or you'll sound like a computer. Also, make numbers as meaningful as possible: For example, instead of saying "in 84.7 percent of cases," say, "in five out of six cases."
- Don't "reach" to interpret the output of statistical modeling. For example, suppose that you have input many variables into an Excel model. You might be able to point out a trend, but you might not be able to say with certainty that if management employs the inputs in the same combination that you used them, they will get exactly the same results.
- Record yourself, if possible, and then evaluate yourself. If that is not possible, have a friend listen to you and evaluate your style. Are you speaking down to your audience? Is your voice unnaturally high-pitched from fear? Are you speaking clearly and distinctly? Is your voice free of distractions, such as "um" and "you know," "uh, so," and "well"?

- If you use a pointer, either a laser pointer or a wand, use it with care. Make sure that you don't accidentally point a laser pointer in someone's face—you'll temporarily blind them. If you're using a wand, don't swing it around or play with it.

Handling Questions

Fielding questions from an audience can be an unpredictable experience because you can't anticipate all the questions that might be asked. When answering questions from an audience, *always treat everyone with courtesy and respect*, no matter what. Use the following strategies to handle questions:

- Anticipate questions. You can gather much of the information that you need as you draft your presentation. Also, if you have a slide that illustrates a key point but doesn't quite fit in your talk, save it—someone might have a question that the slide will answer.
- Mention at the beginning of the talk that you will take questions at the end of your talk. This will (you hope) prevent people from interrupting your presentation. If someone tries to interrupt you, smile and say that you'll be happy to answer all questions when you're finished or that the next graphic will answer their question. (If, however, the person doing the interrupting is the CEO of your company, you want to stop your presentation and answer the question on the spot.)
- When answering a question, first repeat the question if you have *any* doubt that the entire audience might not have heard it. Then deliver the answer to the whole audience, not just the one person who asked the question.
- Be informative and not persuasive, i.e., use facts to answer questions. For instance, if someone asks your opinion about some outcome, you might show an Excel slide that displays the Solver's output, and then you can use that data as the basis for answering the question.
- If you don't know the answer to a question, don't try to fake it. For instance, suppose someone asks you a question about the Scenario Manager that you just can't answer. Be honest. Say, "That is an excellent question but, unfortunately, it's not one that I'm able to answer." At that point, you might ask your instructor whether he or she can answer the question. In a professional setting, you might say that you'll research the answer and e-mail the answer to the person who asked the question.
- Signal when you are finished. You might say, "I have time for one more question." Wrap up the talk yourself.

Handling a "Problem" Audience

A "problem" audience or a heckler is every presenter's nightmare. Fortunately, such experiences are rare. If someone is rude to you or challenges you in a hostile manner, keep cool, be professional, and rely on facts. Know that the rest of the audience sympathizes with your plight and admires your self-control.

The problem that you will most likely encounter is a question from an audience member who lacks technical expertise. For instance, suppose that you explained how to input data into an Access form, but someone didn't understand the explanation that you gave. In such an instance, ask the questioner what part of the explanation is confusing. If you can answer the question briefly, do so. If your answer to the questioner begins to turn into a time-consuming dialogue, offer to give the person one-on-one input later.

Another common problem is someone who asks you a question that you've already answered. The best solution is to answer the question as briefly as possible and use different words (just in case it's the way in which you explained something that confused the person). If the person persists in asking questions that have very obvious answers, either the person is clueless or is trying to heckle you. In that case, you might ask the audience, "Who in the audience would like to answer that question?" The person asking the question will get the hint.

Presentation Toolkit

You can use these forms for preparation, self-analysis, and evaluation of your classmates' presentations (Figures E-24, E-25, and E-26).

Preparation Checklist

Facilities and Equipment

- ❑ The room contains the equipment that I need.
- ❑ The equipment works and I've tested it with my visual aids.
- ❑ Outlets and electrical cords are available and sufficient.
- ❑ All the chairs are aligned so that everyone can see me and hear me.
- ❑ Everyone will be able to see my visual aids.
- ❑ The lights can be dimmed when/if needed.
- ❑ Sufficient light will be available so I can read my notes when the lights are dimmed.

Presentation Materials

- ❑ My notes are available, and I can read them while standing up.
- ❑ My visual aids are assembled in the order that I'll use them.
- ❑ A laser pointer or a wand will be available if needed.

Self

- ❑ I've practiced my delivery.
- ❑ I am comfortable with my presentation and visual aids.
- ❑ I am prepared to answer questions.
- ❑ I can dress appropriate to the situation.

Figure E-24 Preparation Checklist

Tutorial E

Evaluating Access Presentations

Course: _______ **Speaker:** _______________________ **Date:** ______

Rate the presentaton by this criteria:
4=Outstanding 3=Good 2=Adequate 1=Needs Improvement
N/A=Not Applicable

Content

_____ The presentation contained a brief and effective introduction.
_____ Main ideas were easy to follow and understand.
_____ Explanation of database design was clear and logical.
_____ Explanation of using the form was easy to understand.
_____ Explanation of running the queries and their output was clear.
_____ Explanation of the report was clear, logical, and useful.
_____ Additional recommendations for database use were helpful.
_____ Visuals were appropriate for the audience and the task.
_____ Visuals were understandable, visible, and correct.
_____ The conclusion was satisfying and gave a sense of closure.

Delivery

_____ Was poised, confident, and in control of the audience
_____ Made eye contact
_____ Spoke clearly, distinctly, and naturally
_____ Avoided using slang and poor grammar
_____ Avoided distracting mannerisms
_____ Employed natural gestures
_____ Used visual aids with ease
_____ Was courteous and professional when answering questions
_____ Did not exceed time limit

Submitted by: ___________________________

Figure E-25 Form for evaluation of Access presentations

Evaluating Excel Presentations

Course: _______ **Speaker:** _____________________ **Date:** ______

Rate the presentaton by this criteria:
4=Outstanding 3=Good 2=Adequate 1=Needs Improvement
N/A=Not Applicable

Content

_____ The presentation contained a brief and effective introduction.
_____ The explanation of assumptions and goals was clear and logical.
_____ The explanation of software output was logically organized.
_____ The explanation of software output was thorough.
_____ Effective transitions linked main ideas.
_____ Solid facts supported final recommendations.
_____ Visuals were appropriate for the audience and the task.
_____ Visuals were understandable, visible, and correct.
_____ The conclusion was satisfying and gave a sense of closure.

Delivery

_____ Was poised, confident, and in control of the audience
_____ Made eye contact
_____ Spoke clearly, distinctly, and naturally
_____ Avoided using slang and poor grammar
_____ Avoided distracting mannerisms
_____ Employed natural gestures
_____ Used visual aids with ease
_____ Was courteous and professional when answering questions
_____ Did not exceed time limit

Submitted by: _______________________

Figure E-26 Form for evaluation of Excel presentations

Index

A

B

C

D

E

F

G

H

I

J

K

L

Q

R

S

T

U

V

W

Y